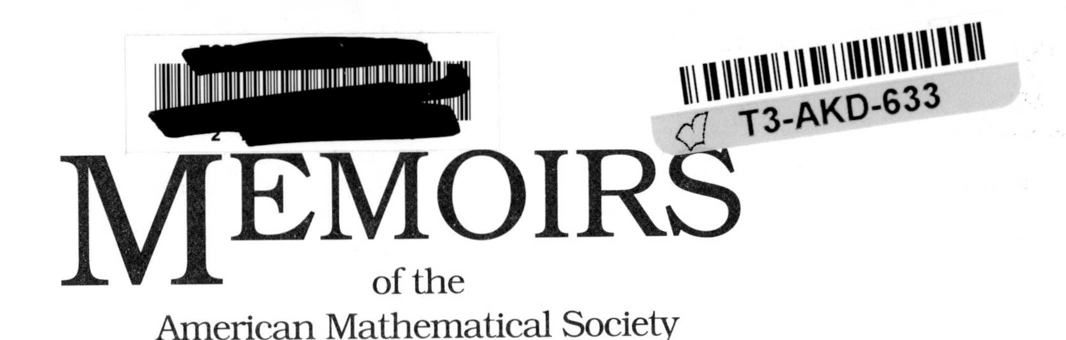

MEMOIRS
of the
American Mathematical Society

Number 445

Atomic Boolean Subspace Lattices and Applications to the Theory of Bases

S. Argyros
M. Lambrou
W. E. Longstaff

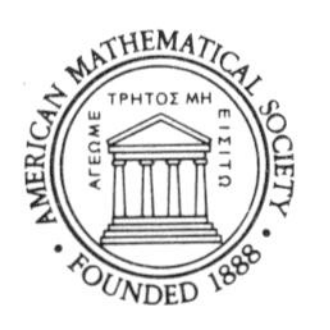

May 1991 • Volume 91 • Number 445 (second of 4 numbers) • ISSN 0065-9266

American Mathematical Society
Providence, Rhode Island

1980 *Mathematics Subject Classification* (1985 *Revision*).
Primary 46B15, 47C05; Secondary 47D30, 06E15.

Library of Congress Cataloging-in-Publication Data

Argyros, S. (Spiros), 1950–
Atomic Boolean subspace lattices and applications to the theory of bases/S. Argyros, M. Lambrou, W. E. Longstaff.
p. cm. – (Memoirs of the American Mathematical Society, ISSN 0065-9266; no. 445)
"May 1991, volume 91, number 445 (second of 4 numbers)."
Includes bibliographical references.
ISBN 0-8218-2511-9
1. Banach spaces. 2. Summability theory. 3. Bases (linear topological spaces) I. Lambrou, M. (Michael), 1952– . II. Longstaff, W. E. (William Ellison), 1945– . III. American Mathematical Society. IV. Title. V. Series.
QA3.A57 no. 445
[QA322.2]
510 s–dc20 90-7545
[515′.732] CIP

Subscriptions and orders for publications of the American Mathematical Society should be addressed to American Mathematical Society, Box 1571, Annex Station, Providence, RI 02901-1571. *All orders must be accompanied by payment.* Other correspondence should be addressed to Box 6248, Providence, RI 02940-6248.

SUBSCRIPTION INFORMATION. The 1991 subscription begins with Number 438 and consists of six mailings, each containing one or more numbers. Subscription prices for 1991 are $270 list, $216 institutional member. A late charge of 10% of the subscription price will be imposed on orders received from nonmembers after January 1 of the subscription year. Subscribers outside the United States and India must pay a postage surcharge of $25; subscribers in India must pay a postage surcharge of $43. Expedited delivery to destinations in North America $30; elsewhere $82. Each number may be ordered separately; *please specify number* when ordering an individual number. For prices and titles of recently released numbers, see the New Publications sections of the NOTICES of the American Mathematical Society.

BACK NUMBER INFORMATION. For back issues see the AMS Catalogue of Publications.

MEMOIRS of the American Mathematical Society (ISSN 0065-9266) is published bimonthly (each volume consisting usually of more than one number) by the American Mathematical Society at 201 Charles Street, Providence, Rhode Island 02904-2213. Second Class postage paid at Providence, Rhode Island 02940-6248. Postmaster: Send address changes to Memoirs of the American Mathematical Society, American Mathematical Society, Box 6248, Providence, RI 02940-6248.

10 9 8 7 6 5 4 3 2 1 95 94 93 92 91

TABLE OF CONTENTS

ABSTRACT

Those families of (closed) subspaces of a (real or complex) Banach space that arise as the set of atoms of an atomic Boolean algebra subspace lattice, abbreviated ABSL, are characterized. This characterization is used to obtain new examples of ABSL's including some with one-dimensional atoms. ABSL's with one-dimensional atoms arise precisely from strong M-bases. The strong rank one density problem for ABSL's is discussed and some affirmative results are presented. Several new areas of investigation in the theory of ABSL's are uncovered.

1. Introduction and preliminaries

Throughout the following 'Banach space' will mean either real or complex Banach space, 'subspace' will mean closed linear subspace and 'operator' will mean bounded linear operator. Also, the letter X denotes a Banach space, the set of operators on X is denoted by $\mathcal{B}(X)$ and the set of subspaces of X is denoted by $\mathcal{C}(X)$.

As usual, for a non-empty subset $\mathcal{L} \subseteq \mathcal{C}(X)$ we let Alg $\mathcal{L}$ denote the set of operators on X which leave every member of $\mathcal{L}$ invariant, that is

$$\text{Alg } \mathcal{L} = \{ T \in \mathcal{B}(X) : T M \subseteq M \text{ , for every } M \in \mathcal{L}\} .$$

Dually, for a non-empty subset $\mathcal{A} \subseteq \mathcal{B}(X)$ we let Lat $\mathcal{A}$ denote the set of invariant subspaces of $\mathcal{A}$, that is

$$\text{Lat } \mathcal{A} = \{ M \in \mathcal{C}(X) : T M \subseteq M \text{ , for every } T \in \mathcal{A}\}.$$

Obviously $\mathcal{L} \subseteq$ Lat Alg $\mathcal{L}$ for any $\mathcal{L} \subseteq \mathcal{C}(X)$. We say that $\mathcal{L}$ is *reflexive* if $\mathcal{L}$ = Lat Alg $\mathcal{L}$. This notion is due to Halmos [9], to whom the dual terminology 'reflexive algebra' is also attributed. Halmos found one of the earliest results in the present context: Every atomic Boolean algebra subspace lattice on complex Hilbert space is reflexive. (Definitions are given below.) Throughout, the abbreviation ABSL is used for atomic Boolean algebra subspace lattice.

Aims of the present work include an investigation of how ABSL's arise, a study of their geometric properties and the provision of new examples. Also, density properties of the finite rank subalgebra of Alg $\mathcal{L}$ are discussed. In the case of one-dimensional atoms, the question of the density of the finite rank subalgebra in the strong operator topology is shown to be

---

Received by editor June 2, 1989

equivalent to a long standing open problem in the theory of bases.

Section 2 includes a characterization (Theorem 2.4) of those families $\{L_\gamma\}_\Gamma$ of non-zero subspaces of X which arise as the set of atoms of some ABSL on X. A necessary and sufficient condition is simply that X be the quasi-direct sum of $\{L_\gamma\}_\Gamma$ in the sense of M.S. Brodskii and G.É. Kisilevskii [3]. For any ABSL $\mathcal{L}$ the set of finite rank operators of Alg $\mathcal{L}$ is shown to be pointwise dense in Alg $\mathcal{L}$ (Corollary 2.6). Also in this section we show how to obtain new ABSL's from old ones.

In Section 3 we elaborate upon the pointwise density property proved in Section 2. For ABSL's with precisely two atoms we show (Theorem 3.1) that this pointwise density property can be improved to density in the strong operator topology. For more general ABSL's this question of strong operator density remains unanswered, but we show that for a certain dense linear manifold of X we get a much stronger conclusion (Theorem 3.5). We also show that this strong operator density property is transmitted in certain cases.

In Section 4, the problem of showing that every ABSL on complex separable Hilbert space has the strong rank one density property is reduced to establishing the property for ABSL's on such spaces satisfying an extreme non-commutative condition, denoted by (G). This is done by using the facts that every commutative ABSL on such a space has the strong rank one density property, and that every ABSL on Hilbert space which is neither commutative nor has property (G) is a 'meshed product' of a commutative one and one with property (G) (Theorem 4.4), and that the meshed product has the density property if and only if each of its 'factors' does (Theorem 4.2). A proof, using the spectral theorem, that ABSL's with precisely two atoms and property (G) have an even stronger density property is included

(Theorem 4.5). This proof is due to K.J. Harrison. For this stronger density property, which we call the metric strong rank one density property, there is a corresponding reduction of the basic problem and corresponding results concerning its transmission. Every ABSL on complex separable Hilbert space with precisely two atoms has the metric strong rank one density property (Theorem 4.6).

In Section 5 the special case where each atom L_γ of an ABSL is one-dimensional is considered. If L_γ is spanned by f_γ, that is if $L_\gamma = \langle f_\gamma \rangle$, then $\{f_\gamma\}_\Gamma$ is a certain generalization of a Schauder basis, called a strong M-basis by Singer in [28] (or a strongly complete family in Markus' terminology [21]; or a 1-series summable family in Ruckle's [27]). Conversely, each strong M-basis $\{f_\gamma\}_\Gamma$ gives rise to an ABSL with atoms $\{\langle f_\gamma \rangle\}_\Gamma$ (Theorem 5.1). This characterization is used to interpret earlier results into the case of one-dimensional atoms (for example, Corollary 5.3). The concepts of strong series summability and series summability [26] are mentioned. The former of these concepts is more stringent than the latter, but not equivalent to it; an example showing this is given in [4]. Because of this example there is an ABSL $\mathcal{L}$ on a Banach space with the metric approximation property for which the unit ball of the finite rank subalgebra fails to be dense in the unit ball of Alg $\mathcal{L}$ in the strong operator topology. We improve the relevant result of [4] somewhat (Example 5.5) by using the powerful constructions of Johnson [14] and Figiel and Johnson [7] to construct a separable Banach space with the approximation property and with a series summable M-basis but which has no strongly series summable M-basis whatsoever. A pointwise boundedness condition is shown to be sufficient for strong operator density of the finite rank subalgebra of an ABSL with one-dimensional atoms (Theorem 5.6). Finally in this section a class of examples of strong M-bases of ℓ^p $(1 \le p < \infty)$ (or C_0) is given. We characterize precisely which members of this class are bases (Theorem 5.8) and, in the case of ℓ^2, we show that no member is similar to an orthonormal basis. For each of these examples the natural projections are strong operator dense (Theorem 5.7) and, although it can be arranged that no subsequence of the sequence of natural projections converge strong operator to the identity operator, some sequence of linear combinations of these projections always does so (Theorem 5.9). So in fact, every member of the class is a strongly series summable M-basis of ℓ^p.

In Section 6 it is shown by an example (Theorem 6.4) that the set of 'slices' of a (finite) ABSL need not be the atoms of an ABSL. This provides a new example of an approximate sum that is not quasi-direct. Also, a 'selection' question is considered and a necessary and sufficient condition for an affirmative answer is given (Theorem 6.2).

Finally in Section 7 we discuss, for an ABSL $\mathcal{L}$, the double commutant property : $(\text{Alg } \mathcal{L})'' = \text{Alg } \mathcal{L}$. Commutative ABSL's (on complex Hilbert space) have this property: since their Alg's are self-adjoint, von Neumann's double commutant theorem applies. For general ABSL's (even on Hilbert space) the equality $(\text{Alg } \mathcal{L})'' = \text{Alg } \mathcal{L}$ may fail. Perhaps surprisingly, we produce an example of an ABSL $\mathcal{L}$ on separable Hilbert space with the double commutant property but whose Alg is as far from being self-ajoint as it can be (Theorem 7.5), inasmuch as every element of $\mathcal{L}$ makes a zero angle with its Boolean complement (with only the trivial exceptions). Also, we provide a sufficient, but not necessary, condition (Corollary 7.2) for the double commutant property. A necessary condition is found (Theorem

7.6(2)), and it is shown to be sufficient for finite ABSL's (Corollary 7.7) but not for infinite ones (Example 7.8).

Most of our results are true for normed spaces without changes in the proofs, but we shall state and prove them for Banach spaces.

The terminology and notation of lattice theory that we use is taken, for the most part, from [2]. Let L be an (abstract) complete lattice and denote the greatest element of L by 1 and the least element by 0. To avoid trivialities we shall assume that $0 \neq 1$. A non-empty subset $A \subseteq L$ is a *complete sublattice* of L if $\vee a_\lambda \in A$ and $\wedge a_\lambda \in A$ for every non-empty family $\{a_\lambda\}$ of elements of A (the join $\vee$ and the meet $\wedge$ being taken in L). Obviously every complete sublattice of L is a complete lattice under the induced partial order. Lattice L is *distributive* if $a\wedge (b\vee c) = (a\wedge b) \vee (a\wedge c)$ and its dual hold, whenever $a,b,c \in L$. Lattice L is *complemented* if, for every $a \in L$, there exists $a' \in L$ such that $a \vee a' = 1$ and $a \wedge a' = 0$. An element $a \in L$ is called an *atom* of L if $a \neq 0$ and $0 \leq b \leq a$, $b \in L$ implies that $b = 0$ or $b = a$. Lattice L is *atomic* if every non-zero element of L contains an atom and is the join of the atoms it contains. Lattice L is a *Boolean algebra* if it is complemented and distributive; in this case, complements are unique and De Morgans' laws hold:

$$(\vee a_\lambda)' = \wedge a'_\lambda \quad \text{and} \quad (\wedge a_\lambda)' = \vee a'_\lambda$$

for every non-empty family $\{a_\lambda\}$ of elements of L. We adopt the convention that $\vee\phi = 0$ and $\wedge\phi = 1$ in any complete lattice.

If L_1 and L_2 are (abstract) complete lattices, a mapping $\varphi : L_1 \to L_2$ is called a *complete homomorphism* if

$$\varphi(\vee a_\lambda) = \vee\varphi(a_\lambda) \quad \text{and} \quad \varphi(\wedge a_\lambda) = \wedge\varphi(a_\lambda)$$

for every non-empty family $\{a_\lambda\}$ of elements of L_1.

Let X be a Banach space. The set $\mathcal{C}(X)$ of all subspaces of X is a complete lattice under the partial order of set-theoretic inclusion; joins in this case are closed linear spans and meets are intersections. A *subspace lattice on* X is a complete sublattice of $\mathcal{C}(X)$ containing the trivial subspaces (0) and X. We repeat that the abbreviation 'ABSL on X' will be used to mean atomic Boolean algebra subspace lattice on X. If $\mathcal{L}$ is an ABSL on X and $\{L_\gamma\}_\Gamma$ is the set of atoms of $\mathcal{L}$, then

$$L'_\gamma = \bigvee_{\mu\neq\gamma} L_\mu \ , \text{ for every } \gamma \in \Gamma \ .$$

It turns out that ABSL's are special types of completely distributive subspace lattices (or strongly reflexive subspace lattices, in the terminology of [19]). We shall not give the definition of complete distributivity here; it can be found in [19]. We merely state the only two properties of them that we shall use. The first is that if $\mathcal{D}$ is a completely distributive subspace lattice on X and $M \in \mathcal{D}$, then (see [15,19])

$$M = \vee\{L \in \mathcal{D} : M \not\subseteq L_-\}$$

where

$$L_- = \vee\{K \in \mathcal{D} : L \not\subseteq K\},$$

for every element $L \in \mathcal{D}$.

The dual of the Banach space X is denoted by X^*. For a non-empty subset $Y \subseteq X$, $Y^\perp$ denotes its annihilator, that is

$$Y^\perp = \{f^* \in X^* : f^*(y) = 0 \ , \text{ for every vector } y \in Y\} \ .$$

In Hilbert space H we identify $Y^{\perp}$ with the set $\{x \in H : (y|x) = 0$, for every vector $y \in Y\}$, where $(\cdot|\cdot)$ denotes the inner-product on H. If $f \in X$, $\langle f \rangle$ denotes the linear span of $\{f\}$. More generally, for a subset $\mathcal{E} \subseteq X$, $\langle \mathcal{E} \rangle$ denotes the linear span of $\mathcal{E}$. If $f^* \in X^*$ and $f \in X$, $f^* \otimes f$ denotes the rank one operator of $\mathcal{B}(X)$ defined by $x \longmapsto f^*(x) f$. In Hilbert space H, if $e, f \in H$, then $e \otimes f$ denotes the operator $x \longmapsto (x|e) f$. If $\{X_j\}_1^{\infty}$ is a sequence of Banach spaces over the same scalar field (real or complex) their ℓ^p-direct sum, for $1 \leq p < \infty$, is denoted by $(\bigoplus_{j=1}^{\infty} X_j)_{\ell^p}$ or by $(X_1 \oplus X_2 \oplus \ldots)_{\ell^p}$, and is the Banach space consisting of those sequences $\{x_j\}_1^{\infty}$ with $x_j \in X_j$ $(j \in \mathbb{Z}^+)$ for which

$$\| \{x_j\}_1^{\infty} \| = (\sum_{j=1}^{\infty} \| x_j \|^p)^{1/p} < \infty .$$

If $X_j = X$ for every $j \in \mathbb{Z}^+$, the same space is denoted by $\ell^p(X)$.

The second property of completely distributive subspace lattices that we shall frequently use is the following special case of a result of [19] (see also [15]).

LEMMA. *If $\mathcal{L}$ is a completely distributive subspace lattice on a Banach space then $f^* \otimes f$ belongs to* Alg $\mathcal{L}$ *if and only if there is a subspace $L \in \mathcal{L}$ such that $f \in L$ and $f^* \in (L_-)^{\perp}$.*

In the above lemma, if $\mathcal{L}$ is an ABSL, the subspace L can be taken to be an atom of $\mathcal{L}$, in which case $L_- = L'$.

A subspace lattice on a complex Hilbert space is *commutative* if the orthogonal projections onto its members pairwise commute.

The *cosine of the angle* between two subspaces M and N of a Hilbert space, satisfying $M \cap N = (0)$, is defined by

$$\cos\,[\text{angle}\,(M,N)] = \sup\{|(x|y)| : x \in M\ ,\ y \in N \text{ and } \|x\| = \|y\| = 1\}\ .$$

It is well known that $M + N$ is a closed sum if and only if this cosine is strictly less than 1 .

A Banach space X is said to have the *approximation property* if, for every compact subset $K \subseteq X$ and every $\epsilon > 0$, there is an operator $F \in \mathcal{B}(X)$ of finite rank such that $\|x - Fx\| < \epsilon$, for every vector $x \in K$ ([18,p.30]).

A Banach space X is said to have the *λ-approximation property* (where $1 \leq \lambda < \infty$) if, for every compact subset $K \subseteq X$ and every $\epsilon > 0$, there is an operator $F \in \mathcal{B}(X)$ of finite rank satisfying $\|F\| \leq \lambda$ such that $\|x - Fx\| < \epsilon$, for every vector $x \in K$. A Banach space X is said to have the *bounded approximation property* if it has the λ-approximation property for some λ , and the *metric approximation property* if it has the 1-approximation property ([18, p.37]).

2. Quasi-direct sums

We begin by characterizing, in two different ways, those families of subspaces of a Banach space X which arise as the set of atoms of an ABSL (Theorems 2.4, 2.5). The first characterization involves the notion, introduced by Brodskii and Kisilevskii [3], of a quasi-direct sum. Let $\{L_\gamma\}_\Gamma$ be a (non-empty) family of (distinct) subspaces of X. The space X is the *approximate sum* of $\{L_\gamma\}_\Gamma$ if $\vee_\Gamma L_\gamma = X$ and

$$(\vee_I L_\gamma) \cap (\vee_J L_\gamma) = (0)$$

for every pair I, J of disjoint subsets of Γ. The space X is the *quasi-direct sum* of $\{L_\gamma\}_\Gamma$ if $\vee_\Gamma L_\gamma = X$ and

$$(\vee_I L_\gamma) \cap (\vee_J L_\gamma) = \vee_{I \cap J} L_\gamma$$

for every pair I, J of subsets of Γ.

Obviously, X is the approximate sum of $\{L_\gamma\}_\Gamma$ if it is their quasi-direct sum. The converse if false. In [3] an example is given of three subspaces of a separable Hilbert space H such that H is their approximate, but not their quasi-direct, sum. In Section 6 we construct three subspaces $L_i (i = 1,2,3)$ such that H is their quasi-direct sum, and three subspaces $K_i \subseteq L_i$ $(i = 1,2,3)$ such that $K_1 \vee K_2 \vee K_3$ is their approximate, but not their quasi-direct, sum. The following geometric property of quasi-direct sums is perhaps of independent interest.

THEOREM 2.1. *If the Banach space X is the quasi-direct sum of the family $\{L_\gamma\}_\Gamma$, then*

$$\cap_\Lambda (\vee_{\Gamma_\lambda} L_\gamma) = \vee\{L_\gamma : \gamma \in \cap_\Lambda \Gamma_\lambda\}$$

for every family $\{\Gamma_\lambda\}_\Lambda$ of subsets of Γ.

PROOF: Clearly

$$\cap_\Lambda(\vee_{\Gamma_\lambda} L_\gamma) \supseteq \vee\{L_\gamma : \gamma \in \cap_\Lambda \Gamma_\lambda\}.$$

Consider the reverse inclusion. For any subset $I \subseteq \cup_\Lambda \Gamma_\lambda$ let I^* denote the set $I \setminus (\cap_\Lambda \Gamma_\lambda)$.

Let $x \in \cap_\Lambda(\vee_{\Gamma_\lambda} L_\gamma)$ be arbitrary. We are to show that $x \in \vee\{L_\gamma : \gamma \in \cap_\Lambda \Gamma_\lambda\}$. Let $\lambda_o \in \Lambda$. Since $x \in \vee_{\Gamma_{\lambda_o}} L_\gamma$ there exists a finite subset I_1 of Γ_{λ_o} and a vector x_1 in the linear span of $\{L_\gamma\}_{I_1}$ such that $\|x - x_1\| < 1$. Since I_1^* is finite, there exists a finite subset $J_1 \subseteq \Lambda$ such that $I_1^* \cap (\cap_{J_1} \Gamma_\mu) = \phi$. For, if $I_1^* = \{\gamma_1, \gamma_2, \ldots, \gamma_k\}$, then for each i with $1 \le i \le k$ there is a $\mu_i \in \Lambda$ such that $\gamma_i \notin \Gamma_{\mu_i}$; so we may take $J_1 = \{\mu_1, \mu_2, \ldots, \mu_k\}$. On the other hand, if I_1^* is empty, any finite subset $J_1 \subseteq \Lambda$ would do. Now

$$x \in \cap_\Lambda(\vee_{\Gamma_\lambda} L_\gamma) \subseteq \cap_{J_1}(\vee_{\Gamma_\lambda} L_\gamma)$$

and so, since we are assuming that X is the quasi-direct sum of the $\{L_\gamma\}_\Gamma$, it follows that

$$x \in \vee\{L_\gamma : \gamma \in \cap_{J_1} \Gamma_\lambda\}.$$

Hence there is a finite subset $I_2 \subseteq \cap_{J_1} \Gamma_\lambda$ and a vector x_2 in the linear span of $\{L_\gamma\}_{I_2}$ such that $\|x - x_2\| < 1/2$. Clearly $I_1^* \cap I_2^* = \phi$.

Next, since $I_1^* \cup I_2^*$ is finite, by an argument similar to the above but with $I_1^* \cup I_2^*$ instead of I_1^*, there exists a finite subset $J_2 \subseteq \Lambda$ such that $(I_1^* \cup I_2^*) \cap (\cap_{J_2} \Gamma_\mu) = \phi$. Since, by quasi-directness again,

$$x \in \vee\{L_\gamma : \gamma \in \cap_{J_2} \Gamma_\lambda \}$$

there exists a finite subset $I_3 \subseteq \cap_{J_2} \Gamma_\lambda$ and a vector x_3 in the linear span of $\{L_\gamma\}_{I_3}$ such that $\|x - x_3\| < 1/3$. Clearly I_1^*, I_2^* and I_3^* are pairwise disjoint. Continuing in this manner we obtain inductively a sequence $\{x_n\}_1^\infty$ of vectors of X and a sequence $\{I_n\}_1^\infty$ of finite subsets of $\cup_\Lambda \Gamma_\lambda$ such that $I_m^* \cap I_n^* = \phi$ if $m \neq n$ and such that, for every $n \in \mathbb{Z}^+$, x_n is in the linear span of $\{L_\gamma\}_{I_n}$ and $\|x - x_n\| < 1/n$.

Put $I = \overset{\infty}{\underset{1}{\cup}} I_{2n}$ and $J = \overset{\infty}{\underset{1}{\cup}} I_{2n-1}$. Then (this is the crucial observation) $I \cap J \subseteq \cap_\Lambda \Gamma_\lambda$ by the pairwise disjointness of the I_n^*'s. As $\lim_n x_{2n} = x = \lim_n x_{2n-1}$ and $x_{2n} \in \vee_I L_\gamma$, $x_{2n-1} \in \vee_J L_\gamma$ we have

$$x \in (\vee_I L_\gamma) \cap (\vee_J L_\gamma) = \vee_{I \cap J} L_\gamma \subseteq \vee\{L_\gamma : \gamma \in \cap_\Lambda \Gamma_\lambda \},$$

and the proof is complete.■

In [3] the authors suggest that the notion of quasi-direct sum can be strengthened by requiring, in addition to $\vee_\Gamma L_\gamma = X$, that

$$\bigcap_{n=1}^\infty (\vee_{\Gamma_n} L_\gamma) = \vee\{L_\gamma : \gamma \in \bigcap_{n=1}^\infty \Gamma_n\}$$

hold for every sequence $\{\Gamma_n\}_1^\infty$ of subsets of Γ. The above theorem shows that the latter conditions are equivalent to quasi-directness.

COROLLARY 2.2. *The Banach space* X *is the quasi-direct sum of the family* $\{L_\gamma\}_\Gamma$ *if and only if* $\cap_I L'_\gamma = \vee_{\Gamma\setminus I} L_\gamma$ *, for every subset* $I \subseteq \Gamma$ *where* $L'_\gamma = \vee_{\mu\neq\gamma} L_\mu$.

PROOF: The sufficiency of the condition follows from the following argument for arbitrary subsets I, J of Γ .

$$\begin{aligned}(\vee_I L_\gamma)\cap(\vee_J L_\gamma) &= (\cap_{\Gamma\setminus I} L'_\gamma) \cap (\cap_{\Gamma\setminus J} L'_\gamma) \\ &= \cap_{\Gamma\setminus(I\cup J)} L'_\gamma \\ &= \vee_{I\cup J} L_\gamma ,\end{aligned}$$

also, taking $I = \phi$ gives $X = \vee_\Gamma L_\gamma$.

The necessity of the condition follows from Theorem 2.1, as follows. For any subset $I \subseteq \Gamma$, take $\Lambda = I$ and $\Gamma_\gamma = \Gamma\setminus\{\gamma\}$. We have $\cap_I L'_\gamma = \cap_I(\vee_{\Gamma_\gamma} L_\mu) = \vee\{L_\mu : \mu \in \cap_I \Gamma_\gamma\} = \vee_{\Gamma\setminus I} L_\gamma$. ∎

The next result is needed in what follows. It's proof is routine and well known and will be omitted.

LEMMA 2.3. *Let* L_1 *and* L_2 *be (abstract) complete lattices and let* $\varphi : L_1 \to L_2$ *be a complete homomorphism. Then* $\varphi(L_1)$ *is a complete sublattice of* L_2 . *If* L_1 *is an atomic Boolean algebra so is* $\varphi(L_1)$ *and the atoms of* $\varphi(L_1)$ *are those elements of the form* $\varphi(a)$ *where* $\varphi(a) \neq \varphi(0)$ *and* a *is an atom of* L_1 .

THEOREM 2.4. *Let* $\{L_\gamma\}_\Gamma$ *be a family of non-zero subspaces of a Banach space* X . *There exists an* ABSL *on* X *having precisely* $\{L_\gamma\}_\Gamma$ *as its set of atoms if and only if* X *is the quasi-direct sum of the* $\{L_\gamma\}_\Gamma$.

PROOF: Suppose that $\{L_\gamma\}_\Gamma$ is the set of atoms of an ABSL $\mathcal{L}$ on X .

As $X \in \mathcal{L}$ we have $X = \vee_\Gamma L_\gamma$. If I and J are subsets of Γ we have

$$(\vee_I L_\gamma) \cap (\vee_J L_\gamma) = \vee\{L_\gamma : L_\gamma \subseteq (\vee_I L_\mu) \cap (\vee_J L_\mu)\}$$
$$= \vee_{I \cap J} L_\gamma ,$$

since $L_\gamma \cap L_\mu = (0)$ if $\gamma \neq \mu$ and $L_\gamma \cap (\vee_K L_\mu) = \vee_K (L_\gamma \cap L_\mu)$ holds for every γ and every subset $K \subseteq \Gamma$ (see Birkhoff [2]).

Conversely, suppose that X is the quasi-direct sum of the $\{L_\gamma\}_\Gamma$. Let $\mathcal{P}$ be the complete atomic Boolean algebra of all subsets of Γ partially ordered by inclusion and define the mapping $\varphi : \mathcal{P} \to \mathcal{C}(X)$ by $\varphi(I) = \vee_I L_\gamma$. It is easily verified, using Theorem 2.1, that φ is a complete homomorphism. The atoms of $\mathcal{P}$ are the singletons and the desired result follows from Lemma 2.3. ∎

The example of Brodskii and Kisilevskii mentioned earlier, of an approximate sum which is not quasi-direct, shows that the sufficiency in Theorem 2.4 cannot be weakened to approximate sums.

The next theorem relates the geometric property of being an ABSL to a pointwise approximation property of the identity operator by certain finite rank operators on X .

THEOREM 2.5. *Let* $\{L_\gamma\}_\Gamma$ *be a family of non-zero subspaces of a Banach space* X . *There exists an* ABSL *on* X *having precisely* $\{L_\gamma\}_\Gamma$ *as its set of atoms if and only if for every* $\epsilon > 0$ *and* $x \in X$ *there exists a finite subset* $I \subseteq \Gamma$, *vectors* f_γ $(\gamma \in I)$ *of* X *and* f^*_γ $(\gamma \in I)$ *of* X^* *with* $f_\gamma \in L_\gamma$, $f^*_\gamma \in (\vee_{\mu \neq \gamma} L_\mu)^\perp$, $(\gamma \in I)$ *such that*

$$\| x - (\Sigma_I f^*_\gamma \otimes f_\gamma) x \| < \epsilon .$$

PROOF: Suppose first that $\{L_\gamma\}_\Gamma$ is the set of atoms of an ABSL $\mathcal{L}$ on X. What we have to prove is a special case of Theorem 3.1 of [15]. A brief proof (adapted for this special case) is included for the convenience of the reader.

Let $\epsilon > 0$ and $x \in X$ be given. We may assume that $x \neq 0$. Put $M = \cap\{L \in \mathcal{L} : x \in L\}$. Then $M \in \mathcal{L}$ and $x \in M$. Now $M = \vee\{L_\gamma : L_\gamma \subseteq M\}$ and if $L_\gamma \subseteq M$ then $x \notin L'_\gamma$ (otherwise we would have $L_\gamma \subseteq M \subseteq L'_\gamma$ and so $L_\gamma \subseteq L'_\gamma$). Thus there exists a finite subset I of $\{\gamma \in \Gamma : L_\gamma \subseteq M\}$ and vectors $f_\gamma \in L_\gamma$ $(\gamma \in I)$ such that $\| x - \Sigma_I f_\gamma \| < \epsilon$. Since for each $\gamma \in I$ we have $x \notin L'_\gamma$, by the Hahn-Banach theorem there exists a vector $f^*_\gamma \in (L'_\gamma)^\perp$ such that $f^*_\gamma(x) = 1$. Thus

$$\| x - (\Sigma_I f^*_\gamma \otimes f_\gamma) x \| < \epsilon .$$

For the converse it is enough, by Theorem 2.4, to show that X is the quasi-direct sum of the $\{L_\gamma\}_\Gamma$. That is, it is enough to show that $X = \vee_\Gamma L_\gamma$ and

$$(\vee_J L_\gamma) \cap (\vee_K L_\gamma) = \vee_{J \cap K} L_\gamma$$

for every pair J, K of subsets of Γ.

For each $x \in X$ and each finite subset $I \subseteq \Gamma$, vectors of the form

$$(\Sigma_I f^*_\gamma \otimes f_\gamma) x = \Sigma_I f^*_\gamma(x) f_\gamma$$

with $f_\gamma \in L_\gamma$ and $f^*_\gamma \in X^*$ $(\gamma \in I)$ belong to the linear span of $\{L_\gamma\}_\Gamma$. By assumption, such vectors approximate x arbitrarily well, so $x \in \vee_\Gamma L_\gamma$. This shows that $X = \vee_\Gamma L_\gamma$. Now let $J, K \subseteq \Gamma$ and let $y \in (\vee_J L_\gamma) \cap (\vee_K L_\gamma)$. Given $\epsilon > 0$ let $I, \{f_\gamma\}_I, \{f^*_\gamma\}_I$ be as in the statement of the theorem (with x replaced by y). If $\gamma \in I$ but

$\gamma \notin J \cap K$ either $\vee_J L_\mu \subseteq \vee_{\mu\neq\gamma} L_\mu$ (if $\gamma \notin J$) or $\vee_K L_\mu \subseteq \vee_{\mu\neq\gamma} L_\mu$ (if $\gamma \notin K$), and so, since $y \in (\vee_J L_\mu) \cap (\vee_K L_\mu)$, $y \in \vee_{\mu\neq\gamma} L_\mu$ and $f_\gamma^*(y) = 0$. Thus the sum

$$(\Sigma_I f_\gamma^* \otimes f_\gamma)y = \Sigma_I f_\gamma^*(y) f_\gamma$$

runs only over indices in $J \cap K$ and hence

$$(\Sigma_I f_\gamma^* \otimes f_\gamma)y \in \vee_{J \cap K} L_\gamma .$$

But as ϵ is arbitrary it follows that $y \in \vee_{J \cap K} L_\gamma$. This shows that

$$(\vee_J L_\gamma) \cap (\vee_K L_\gamma) \subseteq \vee_{J \cap K} L_\gamma$$

and the proof is completed by observing that the reverse inclusion is obvious. ■

If $\mathcal{L}$ is an ABSL on a Banach space X, the lemma in Section 1 says that the operators of the form $\Sigma_I f_\gamma^* \otimes f_\gamma$ appearing in the statement of the preceding theorem are finite rank operators of $\mathrm{Alg}\mathcal{L}$. As these operators form a (two-sided) ideal of $\mathrm{Alg}\mathcal{L}$ we conclude the following.

COROLLARY 2.6. *Let* $\mathcal{L}$ *be an* ABSL *on a Banach space* X. *For every* $x \in X$, $A \in \mathrm{Alg}\mathcal{L}$ *and* $\epsilon > 0$ *there is a finite sum* F *of operators of rank at most one of* $\mathrm{Alg}\mathcal{L}$ *such that* $\|Ax - Fx\| < \epsilon$.

We note that, for any ABSL $\mathcal{L}$ (even on a normed space), the finite rank opeators of $\mathrm{Alg}\mathcal{L}$ are precisely the finite sums of operators of rank at most one of $\mathrm{Alg}\mathcal{L}$ (see [15,20]). This is not always the case for completely distributive subspace lattices, even for those which are also commutative [12].

The pointwise approximation property concluded in the preceding corollary is, of course, weaker than strong (equivalently, weak) operator density of finite sums of operators of rank at most one of $\mathrm{Alg}\mathcal{L}$. It is shown in [15, Theorem 3.1] that a subspace lattice $\mathcal{L}$ (on a normed space) has this pointwise density property if and only if it is completely distributive. If $\mathcal{L}$ is completely distributive are the finite rank operators of Alg$\mathcal{L}$ strong operator dense in Alg$\mathcal{L}$? The answer is not known. Much of what follows is devoted to partial answers to this question for the case when $\mathcal{L}$ is an ABSL.

There is a scarcity of examples of ABSL's. In later sections, we show how application of Theorems 2.4 and 2.5 leads to new examples, but first we present methods (Theorems 2.8 and 2.10) of constructing a new example from an old one. Some more obvious ways of doing this are as follows.

EXAMPLE 2.7.

(1) Let $\{L_\gamma\}_\Gamma$ be the set of atoms of an ABSL on X .

(i) Using Theorem 2.4 it is readily verified that, for every non-empty subset $\Delta \subseteq \Gamma$, $\{L_\gamma\}_\Delta$ is the set of atoms of an ABSL on $\vee_\Delta L_\gamma$.

(ii) Let $\{\Gamma_\lambda\}_\Lambda$ be a family of pairwise disjoint non-empty subsets of Γ with $\cup_\Lambda \Gamma_\lambda = \Gamma$. For each $\lambda \in \Lambda$ put $M_\lambda = \vee_{\Gamma_\lambda} L_\gamma$. Then, using Theorem 2.4, $\{M_\lambda\}_\Lambda$ is the set of atoms of an ABSL on X .

(2) Let $S : X \to Y$ be a bicontinuous linear bijection of the Banach space X onto the Banach space Y . Let $\mathcal{L}$ be an ABSL on X . Then $S\mathcal{L} = \{SL : L \in \mathcal{L}\}$ is easily seen to be an ABSL on Y whose set of atoms is the set of images, under S , of atoms of $\mathcal{L}$.

THEOREM 2.8. *Let* $\{L_\gamma\}_\Gamma$ *be the set of atoms of an* ABSL *on a reflexive Banach space* X . *Then* $\{(L'_\gamma)^\perp\}_\Gamma$ *is the set of atoms of an* ABSL *on* X^* .

PROOF: Let $\mathcal{L}$ denote the ABSL with atoms $\{L_\gamma\}_\Gamma$, and consider the mapping $\varphi : \mathcal{L} \to \mathcal{C}(X^*)$ defined by $\varphi(L) = (L')^\perp$. Since X is reflexive we have $(\vee M_\alpha)^\perp = \cap M_\alpha^\perp$ and $(\cap M_\alpha)^\perp = \vee M_\alpha^\perp$ for every family $\{M_\alpha\}$ of subspaces of X . Using this fact together with De Morgans' laws, it is easily verfied that φ is a complete homomorphism. Now $\varphi(0) = (0)$ and $\varphi(X) = X^*$ so, by Lemma 2.3, $\varphi(\mathcal{L})$ is an ABSL on X^* with atoms $\{\varphi(L_\gamma)\}_\Gamma$, that is $\{(L'_\gamma)^\perp\}_\Gamma$. ∎

The condition that X be reflexive cannot be omitted from the statement of the above theorem. In Section 5 (see the remarks following Corollary 5.3) we give an example where the $\{(L'_\gamma)^\perp\}_\Gamma$ do not even span X^* .

COROLLARY 2.9. *Let $\{L_n\}_1^\infty$ be the set of atoms of an ABSL on a reflexive Banach space X . Every norm bounded sequence $\{x_n\}_1^\infty \subseteq X$ satisfying $x_n \in L_n$ $(n \in \mathbb{Z}^+)$ converges weakly to zero.*

PROOF: If $m \neq n$ then

$$(L'_m)^\perp = (\vee_{i \neq m} L_i)^\perp \subseteq L_n^\perp .$$

Hence for every fixed $m \in \mathbb{Z}^+$ and every $f^* \in (L'_m)^\perp$ we have $f^*(x_n) \to 0$ as $n \to \infty$. Thus $f^*(x_n) \to 0$ for every vector f^* in the linear span of $\{(L'_m)^\perp\}_1^\infty$. By Theorem 2.8 this linear span is dense in X^* . Since also $\{\|x_n\|\}_1^\infty$ is bounded, the result follows. ∎

For Hilbert spaces we have a slightly stronger result than Theorem 2.8.

THEOREM 2.10. *Let $\{L_\gamma\}_\Gamma$ be the set of atoms of an ABSL on a Hilbert space H . For every subset $\Delta \subseteq \Gamma$ the family $\{L_\gamma\}_\Delta \cup \{(L'_\gamma)^\perp\}_{\Gamma \setminus \Delta}$ is also the set of atoms of an ABSL on H .*

PROOF: For $\gamma \in \Gamma$ we have $L'_\gamma \neq H$ so $(L'_\gamma)^\perp \neq (0)$. Also, by De

Morgans' laws

$$\cap_{\Gamma\setminus\Delta} L'_\gamma = (\vee_{\Gamma\setminus\Delta} L_\gamma)' = \vee_\Delta L_\gamma \quad .$$

Hence

$$(\vee_\Delta L_\gamma) \vee (\vee_{\Gamma\setminus\Delta} (L'_\gamma)^\perp) = (\vee_\Delta L_\gamma) \vee (\cap_{\Gamma\setminus\Delta} L'_\gamma)^\perp$$

$$= (\vee_\Delta L_\gamma) \vee (\vee_\Delta L_\gamma)^\perp = H \ .$$

Now, for $\gamma \in \Gamma$ put

$$K_\gamma = \begin{cases} L_\gamma \ , \text{ if } \gamma \in \Delta \ , \\ (L'_\gamma)^\perp \ , \text{ if } \ \gamma \in \Gamma\setminus\Delta \quad . \end{cases}$$

Then $K_\gamma \neq (0)$ and the above shows that $\vee_\Gamma K_\gamma = H$. By Theorem 2.4, to complete the proof it is enough to show that, for every pair I , J of subsets of Γ , we have

$$(\vee_I K_\gamma) \cap (\vee_J K_\gamma) \subseteq \vee_{I \cap J} K_\gamma$$

(once again the reverse inclusion is obvious). Put $I_1 = I \cap \Delta$, $I_2 = I \cap(\Gamma\setminus\Delta)$, $J_1 = J \cap \Delta$ and $J_2 = J \cap(\Gamma\setminus\Delta)$. We have to show that

$$\left[(\vee_{I_1} L_\gamma) \vee (\vee_{I_2} (L'_\gamma)^\perp)\right] \cap \left[(\vee_{J_1} L_\gamma) \vee (\vee_{J_2} (L'_\gamma)^\perp)\right]$$

$$\subseteq (\vee_{I_1 \cap J_1} L_\gamma) \vee (\vee_{I_2 \cap J_2} (L'_\gamma)^\perp). \tag{1}$$

Note that $\gamma \neq \mu$ implies that $L_\gamma \subseteq L'_\mu$. Thus for disjoint subsets Λ_1 and Λ_2 of Γ the subspaces $\vee_{\Lambda_1} L_\gamma$ and $\vee_{\Lambda_2} (L'_\mu)^\perp$ are orthogonal. In particular, $\vee_{I_1} L_\gamma$ and $\vee_{I_2} (L'_\gamma)^\perp$ are orthogonal and so are $\vee_{J_1} L_\gamma$ and $\vee_{J_2} (L'_\gamma)^\perp$.

Let x be a vector belonging to the left hand side of (1). Then $x = y_1 + y_2 = z_1 + z_2$ with

$$y_1 \in V_{I_1} L_\gamma \ , \ y_2 \in V_{I_2}(L'_\gamma)^\perp \ , \ z_1 \in V_{J_1} L_\gamma \ \text{ and } \ z_2 \in V_{J_2}(L'_\gamma)^\perp \ .$$

Thus

$$y_1 - z_1 = z_2 - y_2 \in (V_\Delta L_\gamma) \cap (V_{\Gamma\setminus\Delta}(L'_\gamma)^\perp) = (0) \ .$$

So $x = y + z$ where

$$y = y_1 = z_1 \in (V_{I_1} L_\gamma) \cap (V_{J_1} L_\gamma) = V_{I_1 \cap J_1} L_\gamma$$

and

$$z = y_2 = z_2 \in (V_{I_2}(L'_\gamma)^\perp) \cap (V_{J_2}(L'_\gamma)^\perp) = V_{I_2 \cap J_2}(L'_\gamma)^\perp ,$$

using Theorems 2.4 and 2.8. This completes the proof. ∎

3. Strong rank one density property

A subspace lattice $\mathcal{L}$ on a Banach space X is said to have the *strong rank one density property* if the algebra generated by the operators of Alg $\mathcal{L}$ of rank one is dense in Alg $\mathcal{L}$ in the strong operator topology. The algebra generated by the rank one operators of Alg $\mathcal{L}$ is the set of all finite sums of the form ΣR_j where each R_j is an operator, of rank at most one, belonging to Alg $\mathcal{L}$. Since this algebra is a (two-sided) ideal of Alg $\mathcal{L}$, $\mathcal{L}$ has the strong rank one density property if and only if this algebra contains the identity operator on X in its strong operator closure, that is, if and only if for every $\epsilon > 0$ and every finite set $x_1, x_2, \ldots, x_m$ of vectors of X there exists a finite sum ΣR_j of the form described above such that

$$\|x_i - (\Sigma R_j) x_i\| < \epsilon \text{ , for } i = 1,2,\ldots,m.$$

Which subspace lattices $\mathcal{L}$ have the strong rank one density property? The answer is not known. This question was first raised for Hilbert spaces in [20] where it was shown that such $\mathcal{L}$ are necessarily completely distributive. This result was extended to normed spaces in [15]. In finite-dimensions complete distributivity is also sufficient [20]. On complex, separable Hilbert space totally ordered subspace lattices and, more generally, completely distributive commutative subspace lattices have the strong rank one density property, by the Erdos density theorem [5, Theorem 2] and by [11, Corollary 6.1] and [17, Theorem 3] respectively (actually, Erdos' theorem holds in non-separable spaces as well). Now every ABSL on a Banach space is completely distributive since a result of Tarski [29] says that an (abstract) complete Boolean algebra is completely distributive if and only if it is atomic. Does every ABSL have the strong rank one density property? Corollary 2.6 says that they all have a somewhat weaker density property, but the full answer is still unknown. In fact, a very special case of this question, namely, the case where all the atoms are one-dimensional, was raised by Ruckle in 1974 [27] and is still open (actually, in [27], the question is raised in a different way, but we show that it is an equivalent way).

Much of what follows is concerned with the strong rank one density question for ABSL's. Immediately below we show that the answer is affirmative for ABSL's with two atoms. For complex, separable Hilbert spaces this was first proved by Harrison [10]. His previously unpublished proof, using the spectral theorem, which leads to an even stronger result is postponed to a later section (Theorem 4.5).

THEOREM 3.1. *Every* ABSL *on a Banach space with precisely two atoms has the strong rank one density property.*

PROOF: Let L and M be two non-trivial subspaces of a Banach space X such that $L \cap M = (0)$ and $L \vee M = X$. Let $\mathcal{L}$ be the ABSL with atoms $\{L,M\}$. Let $\epsilon > 0$ and vectors $x_1, x_2, \ldots, x_m$ of X be given. We are to show that there exists a finite sum F of operators of Alg $\mathcal{L}$ of rank at most one such that

$$\|x_i - Fx_i\| < \epsilon \ , \quad \text{for} \quad i = 1,2,\ldots,m \ .$$

Let N be the finite-dimensional subspace spanned by $\{x_1, x_2, \ldots, x_m\}$. Assume for the moment that $L \cap N \neq (0)$ and $M \cap N \neq (0)$ and let $\{y_1, y_2, \ldots, y_k\}$ be a basis of $L \cap N$ and let $\{y_{k+1}, y_{k+2}, \ldots, y_\ell\}$ be a basis of $M \cap N$.

The set $N_o = \{y_1, y_2, \ldots, y_\ell\}$ is linearly independent since if

$$\sum_1^\ell \lambda_i y_i = 0 ,$$

then

$$\sum_1^k \lambda_i y_i = - \sum_{k+1}^\ell \lambda_i y_i .$$

But the vector on the left hand side of the equality belongs to L and the other belongs to M. So each belongs to $L \cap M = (0)$ and so

$$\sum_1^k \lambda_i y_i = - \sum_{k+1}^\ell \lambda_i y_i = 0 .$$

It now follows that $\lambda_i = 0$, for $i = 1,2,\ldots,\ell$.

Now extend N_o to a basis $\{y_1, y_2, \ldots, y_n\}$ of N and assume for the moment that $\ell < n$. Since every x_i is a linear combination of y_i's it is enough to show that, for a suitably adjusted ϵ, denote it by ϵ', there is an operator F as described earlier with

$$\|y_i - Fy_i\| < \epsilon' , \text{ for } i = 1,2,\ldots,n .$$

We build up this F in stages.

First observe that, for every $1 \le i \le k$, y_i does not belong to the span of $\{y_j : 1 \le j \le n , j \ne i\} \cup M$. For, suppose that $1 \le i \le k$ and

$$y_i \in \vee\{y_j : 1 \le j \le n , j \ne i\} \vee M = \vee\{y_j : 1 \le j \le n , j \ne i\} + M .$$

Then, for suitable scalars $\{\mu_j : 1 \le j \le n , j \ne i\}$ and vector $z \in M$ we would have

$$y_i = \sum_{j \ne i} \mu_j y_j + z ,$$

so

$$z = y_i - \sum_{j \ne i} \mu_j y_j \in N .$$

Thus $z \in M \cap N$. Since the latter has basis $\{y_j : k+1 \le j \le \ell\}$, there are scalars $\{\lambda_j : k+1 \le j \le \ell\}$ such that

$$y_i - \sum_{j \ne i} \mu_j y_j = \sum_{k+1}^{\ell} \lambda_j y_j ,$$

and this contradicts the linear independence of $\{y_j : 1 \le j \le n\}$ (note that y_i appears only once).

Hence, by the Hahn-Banach theorem, there exist vectors $\{y_i^* : 1 \le i \le k\} \subseteq X^*$ such that

$$y_i^*(y_i) = 1,\ y_i^*(y_j) = 0 \text{ for every } j \ne i \text{ and } y_i^*(M) = \{0\} .$$

By the lemma in Section 1 the rank one operator $y_i^* \otimes y_i$ $(1 \le i \le k)$ belongs to Alg $\mathcal{L}$.

For each i satisfying $k + 1 \le i \le \ell$ a similar argument produces a vector y_i^* satisfying

$$y_i^*(y_i) = 1 ,\ y_i^*(y_j) = 0 \text{ for every } j \ne i \text{ and } y_i^*(L) = \{0\} ,$$

so again $y_i^* \otimes y_i \in \text{Alg } \mathcal{L}$ $(k+1 \le i \le \ell)$.

For each i satisfying $\ell + 1 \le i \le n$ we do two steps. Firstly, a similar argument to the one immediately above produces vectors y_i^* and $\bar{y}_i^*$ such that

$$y_i^*(y_j) = \delta_{ij} ,\ y_i^*(L) = \{0\} ,$$

and

$$\bar{y}_i^*(y_i) = \delta_{ij} ,\ \bar{y}_i^*(M) = \{0\}$$

(Note that in this case neither $y_i^* \otimes y_i$ nor $\bar{y}_i^* \otimes y_i$ is necessarily in Alg $\mathcal{L}$, so we have to make some adjustments).

Next since $L \vee M = X$ for each i satisfying $\ell + 1 \le i \le n$ there exist vectors $z_i \in L$, $w_i \in M$ such that

$$\| y_i - (z_i + w_i) \| < \epsilon' ,$$

so the operators $y_i^* \otimes w_i$ and $\bar{y}_i^* \otimes z_i$ belong to Alg $\mathcal{L}$. Define F by

$$F = \sum_1^{\ell} y_i^* \otimes y_i + \sum_{\ell+1}^{n} y_i^* \otimes w_i + \sum_{\ell+1}^{n} \bar{y}_i^* \otimes z_i .$$

Then F is a finite sum of rank one operators of Alg $\mathcal{L}$. If we apply F to y_i where $1 \leq i \leq \ell$ all but one of the terms vanish. The non-vanishing one gives $Fy_i = y_i$. On the other hand if we apply F to y_i where $\ell + 1 \leq i \leq n$ all but two terms vanish and we get

$$Fy_i = z_i + w_i .$$

Thus we have

$$\| y_i - Fy_i \| = 0 , \text{ for } 1 \leq i \leq \ell$$

and, for $\ell + 1 \leq i \leq n$,

$$\| y_i - Fy_i \| = \| y_i - (z_i + w_i) \| < \epsilon' .$$

Hence F has the required properties.

If $L \cap N = (0)$ or $M \cap N = (0)$ or $\ell = n$ the desired result is obtained by making the obvious adjustments. This completes the proof.■

Notice that the rank of the operator F produced in the above proof is at most 2m independently of ϵ . So in fact we have proved the following.

COROLLARY 3.2. *If* $\mathcal{L}$ *is an* ABSL *with precisely two atoms on a Banach space* X *and if* $x_1, x_2, \ldots, x_m \in X$, $A \in$ Alg $\mathcal{L}$ *and* $\epsilon > 0$ *are given, then there exists a finite sum* F *of at most* 2m *rank one operators of* Alg $\mathcal{L}$ *such that*

$$\|Ax_i - Fx_i\| < \epsilon , \text{ for } i = 1,2,\ldots,m.$$

Observe that the number 2m in the statement of the preceding corollary cannot be improved upon. For example, in Hilbert space, if $M = L^{\perp}$ and if $e \in L$, $f \in M$ are unit vectors, then it is impossible to approximate $e + f$ to within 1 by a rank one operator of $\mathrm{Alg}\{(0), L, M, H\}$.

Several methods have been mentioned by which a new ABSL can be obtained from an old one. For each of these we show that the strong rank one density property is transmitted to the new if it holds for the old. It is not too difficult to verify this if the new ABSL is obtained by either one of the three ways described in Example 2.7.

THEOREM 3.3. *Let* X *be a reflexive Banach space and let* $\{L_\gamma\}_\Gamma$ *be the set of atoms of an* ABSL *on* X *with the strong rank one density property. Then the* ABSL *on* X^* *with atoms* $\{(L'_\gamma)^{\perp}\}_\Gamma$ *also has the strong rank one density property.*

PROOF: Let $\mathcal{L}$ be the ABSL on X with atoms $\{L_\gamma\}_\Gamma$ and let $\mathcal{L}^*$ be the ABSL on X^* with atoms $\{(L'_\gamma)^{\perp}\}_\Gamma$ (see Theorem 2.8). If $R \in \mathrm{Alg}\,\mathcal{L}$ it is easily verified that $R^*(L'_\gamma)^{\perp} \subseteq (L'_\gamma)^{\perp}$ for every $\gamma \in \Gamma$. So, since every element of $\mathcal{L}^*$ is the closed linear span of $(L'_\gamma)^{\perp}$'s, we have that $R^* \in \mathrm{Alg}\,\mathcal{L}^*$. Additionally, if R has rank at most one so does R^*.

Since the set of finite sums of the form ΣS_j, with S_j an operator of $\mathrm{Alg}\,\mathcal{L}^*$ of rank at most one, is a convex subset of $\mathcal{B}(X^*)$, it is enough to show that the identity operator on X^* belongs to the weak operator closure of this set. So, since X is reflexive, it is enough to show that for every $\epsilon > 0$, every finite set $x_1^*, x_2^*, \ldots, x_m^*$ of vectors of X^* and every set $y_1, y_2, \ldots, y_m$ of vectors of X, there exists a finite sum F of the form ΣS_j with S_j as earlier described, such that

$$|(x_i^* - Fx_i^*)\, y_i| < \epsilon \text{ , for } i = 1,2,\ldots,m \text{ .}$$

Now, by hypothesis, there exists a finite sum G of the form ΣR_i with each $R_i \in \text{Alg } \mathcal{L}$ of rank at most one such that

$$|x_i^*(y_i - Gy_i)| < \epsilon \text{ , for } i = 1,2,\ldots,m \text{ .}$$

But, for each $1 \le i \le m$,

$$|x_i^*(y_i - Gy_i)| = |(x_i^* - G^*x_i^*) y_i|$$

and so we may take $F = G^* = \Sigma R_i^*$. ■

We defer until Section 4 (Theorem 4.3) the proof of the following: If $\{L_\gamma\}_\Gamma$ is the set of atoms of an ABSL with the strong rank one density property on a Hilbert space, then, for every subset $\Delta \subseteq \Gamma$, the ABSL with atoms $\{L_\gamma\}_\Delta \cup \{(L'_\gamma)^\perp\}_{\Gamma\setminus\Delta}$ also has this density property. The one-dimensional version of this result is discussed in Section 5.

We have already remarked that a subspace lattice $\mathcal{L}$ an a Banach space X has the strong rank one density property if and only if for every $\epsilon > 0$ and every finite set $x_1,x_2,\ldots,x_m$ of vectors of X there exists a finite sum F of operators of rank at most one of $\text{Alg } \mathcal{L}$ such that

$$\|x_i - Fx_i\| < \epsilon \text{ , for } i = 1,2,\ldots,m.$$

We have also remarked that this density property implies complete distributivity. The following result says that, in the presence of the latter condition, an operator F of the required form exists provided that the vectors $x_1,x_2,\ldots,x_m$ satisfy a certain strong form of linear independence.

THEOREM 3.4. *Let $\mathcal{L}$ be a completely distributive subspace lattice on a Banach space X . If $x_1,x_2,\ldots,x_m$ are non-zero vectors of X and $M_i \cap M_j = (0)$ whenever $i \ne j$, where $M_i = \cap\{L \in \mathcal{L} : x_i \in L\}$, then for* *every* $\epsilon > 0$ *there exists a finite sum* F *of operators of rank at most one of* $\text{Alg}\,\mathcal{L}$ *such that*

$$\|x_i - Fx_i\| < \epsilon \ , \text{ for } i = 1,2,\ldots,m \ .$$

PROOF: By completeness $M_i \in \mathcal{L}$ and by complete distributivity we have

$$M_i \cap (\bigvee_{j\neq i} M_j) = \bigvee_{j\neq i}(M_i \cap M_j) = (0) \ ,$$

for every $i = 1,2,\ldots,m$. (Since $x_i \in M_i$, this shows in particular that the x_i's are linearly independent.)

Let $\epsilon > 0$ be given. For each i satisfying $1 \le i \le m$, complete distributivity implies (see [19]) that

$$M_i = \vee\{L \in \mathcal{L} : M_i \not\subseteq L_-\}$$

where, for every $L \in \mathcal{L}$, L_- is given by

$$L_- = \vee\{K \in \mathcal{L} : L \not\subseteq K\} \ .$$

Thus, as $x_i \in M_i$, there exist non-zero subspaces $\{L_{ij} : 1 \le j \le n_i\} \subseteq \mathcal{L}$ such that $L_{ij} \subseteq M_i \not\subseteq (L_{ij})_-$, for $j = 1,2,\ldots,n_i$, and there exist vectors $\{f_{ij} : 1 \le j \le n_i\}$ of X such that $f_{ij} \in L_{ij}$ and

$$\|x_i - \sum_{j=1}^{n_i} f_{ij}\| < \epsilon \ .$$

Fix i and j with $1 \le i \le m$ and $1 \le j \le n_i$. As $M_i \not\subseteq (L_{ij})_-$ we also have $x_i \notin (L_{ij})_-$ so, by the Hahn-Banach theorem, there exists a vector $e^*_{ij} \in X^*$ with

$$e^*_{ij}(x_i) = 1 \quad \text{and} \quad e^*_{ij} \in ((L_{ij})_-)^{\perp}.$$

By the lemma in Section 1 the operator $e^*_{ij} \otimes f_{ij}$ belongs to $\text{Alg}\,\mathcal{L}$. Put

$$F = \sum_{i=1}^{m} \sum_{j=1}^{n_i} e^*_{ij} \otimes f_{ij} \ .$$

As $L_{ij} \subseteq M_i$ and $L_{ij} \neq (0)$ we have that

$$L_{ij} \not\subseteq \bigvee_{k\neq i} M_k \ .$$

Thus

$$\bigvee_{k\neq i} M_k \subseteq (L_{ij})_-$$

so $e_{ij}^*(x_k) = 0$ for $k \neq i$. Hence for $1 \leq k \leq m$ we have

$$Fx_k = \sum_{i=1}^{m} \sum_{j=1}^{n_i} e_{ij}^*(x_k) f_{ij} = \sum_{j=1}^{n_k} e_{kj}^*(x_k) f_{kj} = \sum_{j=1}^{n_k} f_{kj} ,$$

so $\|x_k - Fx_k\| < \epsilon$. This completes the proof. ∎

If, for an ABSL, we impose the condition on $x_1, x_2, \ldots, x_m$ that they be vectors in the (dense) linear span of its atoms, the next theorem tells us that we can get a very strong conclusion. Notice however that this result does not imply the strong rank one density property as the proof does not provide a control on the norm of F.

THEOREM 3.5. *Let* $\mathcal{L}$ *be an ABSL on a Banach space* X. *For every finite set* $x_1, x_2, \ldots, x_m$ *of vectors of* X, *each of which belongs to the (dense) linear span of the set of atoms of* $\mathcal{L}$, *there exists a finite sum* F *of operators of rank at most one of* Alg $\mathcal{L}$ *such that*

$$Fx_i = x_i , \text{ for } i = 1,2,\ldots,m .$$

PROOF: We may suppose that $x_i \neq 0$ $(1 \leq i \leq m)$. By hypothesis there exist distinct atoms $K_1, K_2, \ldots, K_n$ of $\mathcal{L}$ and vectors $\{x_{ij} : 1 \leq i \leq m\} \subseteq K_j$, for $j = 1,2,\ldots,n$, such that $x_i = \sum_{j=1}^{n} x_{ij}$ and not every element of $\{x_{ij} : 1 \leq i \leq m\}$ is zero. For each j satisfying $1 \leq j \leq n$ there exists a non-empty subset $\Delta_j \subseteq \{1,2,\ldots,m\}$ such that $\{x_{kj} : k \in \Delta_j\}$ is a basis for the linear span of $\{x_{1j}, x_{2j}, \ldots, x_{mj}\}$. Let $1 \leq j \leq n$ and let $k \in \Delta_j$. Since $K_j \cap K_j' = (0)$ it follows that

$$x_{kj} \notin \langle \{x_{ij} : i \in \Delta_j \text{ and } i \neq k\} \rangle \vee K_j' .$$

Hence, by the Hahn-Banach theorem, there exists a vector $x_{kj}^* \in X^*$ such that

$$x_{kj}^*(x_{kj}) = 1 \text{ and } \langle \{x_{ij} : i \in \Delta_j \text{ and } i \neq k\} \rangle \vee K_j' \subseteq \ker x_{kj}^* .$$

By the lemma in Section 1, since $x_{kj} \in K_j$ and $x^*_{kj} \in (K'_j)^{\perp}$, we have that $x^*_{kj} \otimes x_{kj} \in \operatorname{Alg} \mathcal{L}$.

Put

$$F_j = \Sigma_{k \in \Delta_j} x^*_{kj} \otimes x_{kj} .$$

If $p \in \Delta_j$ clearly $F_j(x_{pj}) = x_{pj}$. By linear dependence we also have that

$$F_j(x_{pj}) = x_{pj} , \text{ for } p \notin \Delta_j \text{ with } 1 \le p \le m .$$

Also, if $q \neq j$ and $1 \le q \le n$, $1 \le r \le m$, we have $F_j(x_{rq}) = 0$ since $x_{rq} \in K_q \subseteq K'_j$.

Put $F = \sum_{j=1}^{n} F_j$. Then, for $i = 1,2,\ldots,m$ we have

$$Fx_i = F(\sum_{t=1}^{n} x_{it}) = \sum_{t=1}^{n} \sum_{j=1}^{n} F_j(x_{it}) = \sum_{j=1}^{n} x_{ij} = x_i ,$$

and the proof is complete. ■

4. Meshed products

In this section we work, for the most part, in real or complex, non-zero, not necessarily separable Hilbert space H. It will be made clear if restriction to separable and complex space is intended. We are concerned primarily with the question of strong rank one density of ABSL's on such a Hilbert space H. It has already been mentioned that every commutative ABSL (and, more generally, every commutative completely distributive subspace lattice) on a complex, separable Hilbert space has the strong rank one density property. Now, an ABSL on such a space is commutative if and only if

$$K \cap (K')^{\perp} = K \text{ , for every atom } K \text{ .}$$

At the other extreme there are ABSL's with the following property

$$(G) \quad K \cap (K')^{\perp} = (0) \text{ , for every atom } K \text{ .}$$

We show (Theorem 4.4) that every ABSL on H which is neither commutative nor has property (G) is a 'meshed product' of a commutative one and one with property (G) and thereby reduce the problem of proving that every ABSL on separable complex Hilbert space has the strong rank one density property to proving that every ABSL (on such a space) with property (G) has it.

Combining some earlier remarks, an ABSL $\mathcal{L}$ on a (real or complex) Banach space (even a normed space) has the strong rank one density property if and only if the identity operator belongs to the strong operator closure of the set of finite rank operators of Alg $\mathcal{L}$. We shall use this fact in what follows. First, however, some preliminary results.

THEOREM 4.1. *Let* $H_i (i = 1,2)$ *be a (non-zero) Hilbert space and let* $\mathcal{L}_i (i=1,2)$ *be an* ABSL *on* H_i . *Let* $\{K_\alpha\}_{\Gamma_1}$ *(respectively,* $\{L_\beta\}_{\Gamma_2}$ *) denote the set of atoms of* $\mathcal{L}_1$, *(respectively,* $\mathcal{L}_2$*)* . *Let* $\overline{\Gamma}_i$ $(i=1,2)$ *be a set satisfying* $\Gamma_i \subseteq \overline{\Gamma}_i$ *with* $|\overline{\Gamma}_1| = |\overline{\Gamma}_2|$, *and let* $\theta : \overline{\Gamma}_1 \to \overline{\Gamma}_2$ *be a bijection. For* $\alpha \in \overline{\Gamma}_1 \setminus \Gamma_1$ *define* $K_\alpha = (0)$ *and for* $\beta \in \overline{\Gamma}_2 \setminus \Gamma_2$ *define* $L_\beta = (0)$. *The set* $\mathcal{L}$ *of subspaces of* $H_1 \oplus H_2$ *of the form*

$$(\vee\{K_\alpha : \alpha \in I\}) \oplus (\vee\{L_\beta : \beta \in \theta(I)\})$$

where I *is an arbitrary subset of* $\overline{\Gamma}_1$, *is an* ABSL *on* $H_1 \oplus H_2$. *The atoms of* $\mathcal{L}$ *are those non-zero subspaces of the form*

$$K_\alpha \oplus L_{\theta(\alpha)} , \text{ with } \alpha \in \overline{\Gamma}_1 .$$

PROOF: Consider the mapping $\varphi : \mathcal{P}(\overline{\Gamma}_1) \to \mathcal{C}(H_1 \oplus H_2)$, from the power set of $\overline{\Gamma}_1$ to the set of subspaces of $H_1 \oplus H_2$, defined by

$$\varphi(I) = (\vee\{K_\alpha : \alpha \in I\}) \oplus (\vee\{L_\beta : \beta \in \theta(I)\}) .$$

If $\{M_\lambda\}_J$ (respectively, $\{N_\lambda\}_J$) is a family of subspaces of H_1 (respectively, of H_2) it is not difficult to verify that

$$\vee_J(M_\lambda \oplus N_\lambda) = (\vee_J M_\lambda) \oplus (\vee_J N_\lambda)$$

and

$$\cap_J(M_\lambda \oplus N_\lambda) = (\cap_J M_\lambda) \oplus (\cap_J N_\lambda) .$$

Using this it follows that φ is a complete homomorphism, with domain the complete atomic Boolean algebra $\mathcal{P}(\overline{\Gamma}_1)$. Since $\mathcal{L} = \varphi(\mathcal{P}(\overline{\Gamma}_1))$ and $\varphi(\phi) = (0)$, $\varphi(\overline{\Gamma}_1) = H_1 \oplus H_2$ the desired results follow from Lemma 2.3. ∎

We will say that the ABSL $\mathcal{L}$ on $H_1 \oplus H_2$, as in the statement of the preceding theorem, is a *meshed product* of the ABSL's $\mathcal{L}_1$ (on H_1) and $\mathcal{L}_2$ (on H_2) .

THEOREM 4.2. *With* H_i , $\mathcal{L}_i$ $(i = 1,2)$ *and* $\mathcal{L}$ *as in the statement of Theorem* 4.1 , $\mathcal{L}$ *has the strong rank one density property if and only if each of* $\mathcal{L}_1$ *and* $\mathcal{L}_2$ *does* .

PROOF: Every operator T on $H_1 \oplus H_2$ has a matrix representation

$$\begin{bmatrix} T_{11} & T_{12} \\ T_{21} & T_{22} \end{bmatrix}$$

with T_{ij} an operator from H_j to H_i . Clearly, T has finite rank if and only if each T_{ij} does. An operator leaves every element of an ABSL invariant if and only if it leaves every atom invariant. Using these facts it is easily shown that the set of finite rank operators of Alg $\mathcal{L}$ consists of those operators T for which T_{ii} $(i = 1,2)$ is a finite rank operator of Alg $\mathcal{L}_i$ and T_{12} , T_{21} are finite rank operators satisfying

$$T_{12}\, L_{\theta(\alpha)} \subseteq K_\alpha \text{ and } T_{21}\, K_\alpha \subseteq L_{\theta(\alpha)} \text{ , for every } \alpha \in \bar{\Gamma}_1 \text{ .}$$

The desired result now follows easily once one observes that a net

$$\begin{bmatrix} T^\gamma_{11} & T^\gamma_{12} \\ T^\gamma_{21} & T^\gamma_{22} \end{bmatrix}$$

converges to the identity operator in the strong operator topology if and only if each of the nets $\{T^\gamma_{11}\}$, $\{T^\gamma_{22}\}$ do and both the nets $\{T^\gamma_{12}\}$, $\{T^\gamma_{21}\}$ converge strong operator to zero. ■

Now, as promised in Section 3, we prove a stronger result than Theorem 3.3 if the underlying space is a Hilbert space (see Theorem 2.10).

THEOREM 4.3. *Let* H *be a (non-zero) Hilbert space and let* $\{L_\gamma\}_\Gamma$ *be the set of atoms of an* ABSL *on* H *with the strong rank one density property. Then, for every subset* $\Delta \subseteq \Gamma$, *the* ABSL *on* H *with atoms* $\{L_\gamma\}_\Delta \cup \{(L'_\gamma)^\perp\}_{\Gamma\setminus\Delta}$ *also has the strong rank one density property.*

PROOF: Let $\mathcal{L}_\Delta$ be the ABSL on H with atoms $\{L_\gamma\}_\Delta \cup \{(L'_\gamma)^\perp\}_{\Gamma\setminus\Delta}$. By Theorem 3.3 we may assume that both the subsets $\Gamma_1 = \Delta$ and $\Gamma_2 = \Gamma\setminus\Delta$ of Γ are non-empty. Then both $H_1 = \vee_{\Gamma_1} L_\gamma$ and $H_2 = \vee_{\Gamma_2} (L'_\gamma)^\perp$ are non-zero. Now (see Example 2.7) $\{L_\gamma\}_{\Gamma_1}$ is the set of atoms of an ABSL, $\mathcal{L}_1$ say, on H_1 and $\{(L'_\gamma)^\perp\}_{\Gamma_2}$ is the set of atoms of an ABSL, $\mathcal{L}_2$ say, on H_2 . Also, by Theorem 3.3 and earlier remarks, both $\mathcal{L}_1$ and $\mathcal{L}_2$ have the strong rank one density property. Now $H_2 = H_1^\perp$ so if we take, as in the statement of Theorem 4.1, $\overline{\Gamma}_1 = \overline{\Gamma}_2 = \Gamma$ and θ the identity mapping, then the corresponding meshed product of $\mathcal{L}_1$ and $\mathcal{L}_2$ is unitarily equivalent to $\mathcal{L}_\Delta$ (by the unitary mapping $U : H_1 \oplus H_2 \to H$ defined by $U(x,y) = x + y$) . By Theorem 4.2 this meshed product has the strong rank one density property and hence so does $\mathcal{L}_\Delta$. ■

In the proof of the following theorem recall that, for every atom K of an ABSL $\mathcal{L}$,

$$K' = \vee\{L \in \mathcal{L} : L \text{ is an atom and } L \neq K\} .$$

THEOREM 4.4. *Let* $\mathcal{L}$ *be an* ABSL *on a Hilbert space* H . *Suppose that* $\mathcal{L}$ *is neither commutative nor has property* (G) . *Then there exist non-zero Hilbert spaces* H_1 *and* H_2 , *over the same scalar field as* H , *and* ABSL's $\mathcal{L}_1$ *on* H_1 *and* $\mathcal{L}_2$ *on* H_2 *such that* $\mathcal{L}_1$ *is commutative,* $\mathcal{L}_2$ *has property* (G) *and* $\mathcal{L}$ *is unitarily equivalent to a meshed product of* $\mathcal{L}_1$ *and* $\mathcal{L}_2$.

PROOF: Let $\{K_\lambda\}_\Lambda$ denote the set of atoms of $\mathcal{L}$. Since $\mathcal{L}$ is not commutative, $|\Lambda| \geq 2$. For each $\lambda \in \Lambda$ put

$$M_\lambda = K_\lambda \cap (K'_\lambda)^\perp .$$

Since $K_\mu \subseteq K'_\lambda$ if $\mu \neq \lambda$, the subspaces $\{M_\lambda\}_\Lambda$ are pairwise orthogonal. Put $M = \vee_\Lambda M_\lambda$ and put $N = M^\perp$. For each $\lambda \in \Lambda$ put $N_\lambda = N \cap K_\lambda$. Then

$$N_\lambda^\perp = N^\perp \vee K_\lambda^\perp = M \vee K_\lambda^\perp = M_\lambda \vee K_\lambda^\perp = (K_\lambda \cap M_\lambda^\perp)^\perp$$

(using $M_\mu \subseteq K_\lambda^\perp$ if $\mu \neq \lambda$). Thus $N_\lambda = K_\lambda \cap M_\lambda^\perp = K_\lambda \ominus M_\lambda$ and so $K_\lambda = M_\lambda \oplus N_\lambda$ (that is, each atom 'splits'). Next we show that

$$N \cap (\vee_\Sigma K_\lambda) = \vee_\Sigma (N \cap K_\lambda) \text{ , for every non-empty subset } \Sigma \subseteq \Lambda .$$

We have

$$\begin{aligned} N \cap (\vee_\Sigma K_\lambda) &= N \cap [(\vee_\Sigma M_\lambda) \oplus (\vee_\Sigma N_\lambda)] \\ &= (\vee_\Sigma N_\lambda) \vee (N \cap (\vee_\Sigma M_\lambda)) \\ &= \vee_\Sigma N_\lambda \\ &= \vee_\Sigma (N \cap K_\lambda) . \end{aligned}$$

From this it follows that the mapping $\varphi : \mathcal{L} \to \mathcal{C}(N)$ defined by $\varphi(L) = N \cap L$ is a complete homomorphism. Since $\varphi(0) = (0)$, by Lemma 2.3, $\mathcal{L}_2 = \varphi(\mathcal{L})$ is an ABSL on N with atoms $\{N_\lambda : N_\lambda \neq (0)\}$. Let $L \in \mathcal{L}$ be non-zero and put $\Sigma = \{\lambda \in \Lambda : K_\lambda \subseteq L\}$. Then $\varphi(L) = \vee_\Sigma N_\lambda$ since $L = \vee_\Sigma K_\lambda$. Since

$$L = \vee_\Sigma K_\lambda = (\vee_\Sigma M_\lambda) \oplus (\vee_\Sigma N_\lambda) ,$$

it follows that $L \ominus \varphi(L) = \vee_\Sigma M_\lambda$. It is now fairly easy to show that the mapping $\psi : \mathcal{L} \to \mathcal{C}(M)$ defined by $\psi(L) = L \ominus \varphi(L)$ is a complete homomorphism. Since $\psi(0) = (0)$, by Lemma 2.3, $\mathcal{L}_1 = \psi(\mathcal{L})$ is an ABSL on M with atoms $\{M_\lambda : M_\lambda \neq (0)\}$. Since the atoms of $\mathcal{L}_1$ are pairwise orthogonal, $\mathcal{L}_1$ is commutative. Next we show that $\mathcal{L}_2$ has property (G). Note that since $\mathcal{L}$ is not commutative, $\mathcal{L}_2$ has at least two atoms. In particular $N \neq (0)$. We show that for every $\lambda \in \Lambda$,

$$N_\lambda \cap (N \ominus (\bigvee_{\mu\neq\lambda} N_\mu)) = (0) .$$

Let $\lambda \in \Lambda$. Then

$$\begin{aligned} N_\lambda \cap (N \ominus (\bigvee_{\mu\neq\lambda} N_\mu)) &= N_\lambda \cap N \cap (\bigvee_{\mu\neq\lambda} N_\mu)^\perp \\ &= N_\lambda \cap (\bigcap_{\mu\neq\lambda} N_\mu^\perp) \\ &= N_\lambda \cap (\bigcap_{\mu\neq\lambda} (K_\mu^\perp \oplus M_\mu)) . \end{aligned}$$

For every $\mu \neq \lambda$,

$$N_\lambda \cap (K_\mu^\perp \oplus M_\mu) \subseteq M_\mu^\perp \cap (K_\mu^\perp \oplus M_\mu) = K_\mu^\perp .$$

Thus

$$\begin{aligned} N_\lambda \cap (N \ominus (\bigvee_{\mu\neq\lambda} N_\mu)) &\subseteq (\bigcap_{\mu\neq\lambda} K_\mu^\perp) \cap N_\lambda \\ &= (\bigvee_{\mu\neq\lambda} K_\mu)^\perp \cap K_\lambda \cap M_\lambda^\perp \\ &= M_\lambda \cap M_\lambda^\perp \\ &= (0) , \end{aligned}$$

and so

$$N_\lambda \cap (N \ominus (\bigvee_{\mu\neq\lambda} N_\mu)) = (0) .$$

It follows that $\mathcal{L}_2$ has property (G) . Since $\mathcal{L}$ does not have property (G) , $M \neq (0)$. Now put $H_1 = M$, $H_2 = N$, $\Gamma_1 = \{\lambda \in \Lambda : M_\lambda \neq (0)\}$, $\Gamma_2 = \{\lambda \in \Lambda : N_\lambda \neq (0)\}$, $\overline{\Gamma}_1 = \overline{\Gamma}_2 = \Lambda$ and let $\theta : \overline{\Gamma}_1 \to \overline{\Gamma}_2$ be the identity mapping. Let $U : H_1 \oplus H_2 \to H$ be the unitary operator given by $U(x,y) = x + y$. Then the image under U of the corresponding meshed product of $\mathcal{L}_1$ on H_1 and $\mathcal{L}_2$ on H_2 is obviously $\mathcal{L}$. This completes the proof. ■

In view of the preceding theorem and Theorem 4.2 and the fact that commutative ABSL's on separable complex Hilbert spaces have the strong rank one density property, the problem of which ABSL's on such spaces have this density property reduces to consideration of ABSL's having property (G) .

For ABSL's with two atoms the following result is due to Harrison [10] (cf Theorem 3.1). The authors thank Dr Harrison for his permission to include his proof in this paper.

THEOREM 4.5. (K.J. Harrison) *Let* $\mathcal{L}$ *be an* ABSL *on a complex separable Hilbert space* H *having precisely two atoms and having property* (G) . *The identity operator on* H *is the limit, in the strong operator topology, of a sequence of finite rank operators in the unit ball of* Alg $\mathcal{L}$.

PROOF: Let K_1 and K_2 be the atoms of $\mathcal{L}$. Then $K_1 \cap K_2$, $K_1^{\perp} \cap K_2$, $K_1 \cap K_2^{\perp}$ and $K_1^{\perp} \cap K_2^{\perp}$ all equal (0) (that is, K_1 and K_2 are in *generic position*). By a result of Halmos [8] , $\mathcal{L}$ is unitarily equivalent to

$$\mathcal{L}_o = \{(0),\ G(T),\ G(-T),\ K \oplus K\}$$

for some complex separable Hilbert space K and some positive contraction $T \in \mathcal{B}(K)$ satisfying $\ker T = \ker(I - T) = (0)$. If $U : H \to K \oplus K$ is the implementing unitary then, since Alg $\mathcal{L} = U^{-1}(\text{Alg } \mathcal{L}_o)\, U$, it is enough to prove the result for $\mathcal{L}_o$.

It is easily verified that for every pair A,B of finite rank operators on K satisfying TA = BT , the operator

$$\begin{bmatrix} A & 0 \\ 0 & B \end{bmatrix}$$

is a finite rank operator in Alg $\mathcal{L}_o$.

Let $T = \int_0^1 t\, dE(t)$ be the spectral representation of T. Since T is a positive injective contraction, $E(0) = 0$ and $E(1) = I$. For every $n \in \mathbb{Z}^+$ let

$$\mathcal{P}_n : 0 = t_0^{(n)} < t_1^{(n)} < t_2^{(n)} < \ldots < t_{2^n}^{(n)} = 1$$

be the corresponding dyadic partition of $[0,1]$: $t_j^{(n)} - t_{j-1}^{(n)} = 2^{-n}$ for every j. For every j satisfying $1 \le j \le 2^n$ define

$$\delta E_j^{(n)} = E(t_j^{(n)}) - E(t_{j-1}^{(n)})$$

and

$$\delta T_j^{(n)} = T(\delta E_j^{(n)}) = (\delta E_j^{(n)})T .$$

Then

$$I = \sum_{j=1}^{2^n} \delta E_j^{(n)} \quad \text{and} \quad T = \sum_{j=1}^{2^n} \delta T_j^{(n)} ,$$

and if $j \neq k$,

$$(\delta T_j^{(n)})\,(\delta T_k^{(n)}) = (\delta E_j^{(n)})\,(\delta E_k^{(n)}) = 0 .$$

For every $n \in \mathbb{Z}^+$ define

$$C^{(n)} = \sum_{j=1}^{2^n} \delta T_j^{(n)} / t_j^{(n)} .$$

Since

$$\delta T_j^{(n)} \le t_j^{(n)}\, \delta E_j^{(n)} ,$$

we have

$$0 \le \delta T_j^{(n)} / t_j^{(n)} \le \delta E_j^{(n)}$$

and so $0 \le C^{(n)} \le I$. In particular, $\| C^{(n)} \| \le 1$.

In fact we can write

$$C^{(n)} = \int_0^1 \phi^{(n)}(t)\, dE(t)$$

where

$$\phi^{(n)}(t) = t/t_j^{(n)} \ , \ \text{for} \ \ t_{j-1}^{(n)} < t \le t_j^{(n)} \ ,$$

and $\phi^{(n)}(0) = 0$. Since for each $t \in (0,1]$, $\phi^{(n)}(t) \uparrow 1$ as $n \to \infty$ and since $E(0) = 0$, $C^{(n)} \to I$ in the strong operator topology.

Let $\{x_n\}_1^\infty$ be a sequence of vectors which is dense in K . For each $n \in \mathbb{Z}^+$ let $M^{(n)}$ denote the subspace of K spanned by the vectors

$$\delta E_j^{(n)} x_k \ , \ \text{for } j = 1,2,\ldots,2^n \ \text{ and } \ k = 1,2,\ldots,n \ .$$

Let $P^{(n)}$ denote the orthogonal projection onto $M^{(n)}$. Then

(i) each $M^{(n)}$ is finite-dimensional, so each $P^{(n)}$ has finite rank,

(ii) $M^{(n)} \subseteq M^{(n+1)}$ since

$$\delta E_j^{(n)} = \delta E_{2j-1}^{(n+1)} + \delta E_{2j}^{(n+1)} \ ,$$

and (iii) $P^{(n)} \to I$ in the strong operator topology since

$$x_k = \sum_{j=1}^{2^n} \delta E_j^{(n)} x_k \ , \ \text{for each } \ k \in \mathbb{Z}^+ \ .$$

For $1 \le j \le 2^n$ let $\delta M_j^{(n)}$ denote the subspace spanned by the vectors

$$\delta E_j^{(n)} x_k \ , \ \text{for } \ k = 1,2,\ldots,n$$

and let $\delta P_j^{(n)}$ denote the corresponding orthogonal projection.

Then $P^{(n)} = \sum_{j=1}^{2^n} \delta P_j^{(n)}$ and

$$\delta P_j^{(n)} = P^{(n)}(\delta E_j^{(n)}) = (\delta E_j^{(n)})\, P^{(n)} \ .$$

For each $n \in \mathbb{Z}^+$ define $A_n = P^{(n)}C^{(n)}$ and $B_n = C^{(n)}P^{(n)}$. Since both $\{P^{(n)}\}$ and $\{C^{(n)}\}$ converge to the identity operator in the strong operator topology so do $\{A_n\}$ and $\{B_n\}$. Furthermore, since each

$P^{(n)}$ has finite rank, so does each of A_n and B_n. Also, $\| A_n \| \le 1$ and $\| B_n \| \le 1$. We have

$$T A_n = TP^{(n)}C^{(n)} = \sum_{j=1}^{2^n} T P^{(n)}(\delta T_j^{(n)}) / t_j^{(n)}$$

$$= \sum_{j=1}^{2^n} T P^{(n)}(\delta E_j^{(n)}) T / t_j^{(n)}$$

$$= \sum_{j=1}^{2^n} T(\delta P_j^{(n)}) T / t_j^{(n)}$$

and similarly

$$B_n T = C^{(n)}P^{(n)} T = \sum_{j=1}^{2^n} T(\delta P_j^{(n)}) T / t_j^{(n)} .$$

Thus $TA_n = B_n T$ for every $n \in \mathbb{Z}^+$ and so

$$\begin{bmatrix} A_n & 0 \\ 0 & B_n \end{bmatrix}$$

is a finite rank operator in $\text{Alg}\mathcal{L}_o$, of norm at most 1.

Obviously,

$$\begin{bmatrix} A_n & 0 \\ 0 & B_n \end{bmatrix} \to \begin{bmatrix} I & 0 \\ 0 & I \end{bmatrix}$$

in the strong operator topology. This completes the proof. ■

If $\mathcal{L}$ is as in the statement of the preceding theorem, then $\mathcal{L}$ has a density property stronger than the strong rank one density property; it has what we shall call the metric strong rank one density property. In general, a subspace lattice $\mathcal{L}$ on a Banach space X is said to have the *metric strong rank one density property* if the unit ball of the algebra generated by the operators of Alg $\mathcal{L}$ of rank one is dense in the unit ball of Alg $\mathcal{L}$ in the strong operator topology. Arguing as before, $\mathcal{L}$ has this density property if and only if the identity operator belongs to the strong operator closure of the unit ball of the algebra generated by the rank one operators of Alg $\mathcal{L}$, that is, if and only if for every $\epsilon > 0$ and every finite set $x_1, x_2, \ldots, x_m$ of vectors of X there exists a finite sum ΣR_j, with each $R_j \in$ Alg $\mathcal{L}$ of rank at most one, such that

$$\| \Sigma R_j \| \le 1 \quad \text{and} \quad \| x_i - (\Sigma R_j) x_i \| < \epsilon , \quad \text{for} \quad i = 1,2,\ldots,m .$$

Clearly, if X does not have the metric approximation property then no subspace lattice on it can have the metric strong rank one density property. Which (necessarily completely distributive) subspace lattices (on spaces with the metric approximation property) have the metric strong rank one density property? The answer is unknown although there are several partial answers.

In [5, Theorem 3] Erdos shows that totally ordered subspace lattices (nests) on complex Hilbert spaces have this property. Minor modifications to the final proof of [5] show that if a subspace lattice $\mathcal{L}$ on a Banach space X has the metric strong rank one density property, then the norm closure of the algebra generated by the rank one operators of Alg $\mathcal{L}$ is the set of compact operators of Alg $\mathcal{L}$.

Not every ABSL on a Banach space (with the metric approximation property) has the metric strong rank one density property by a remarkable example of Crone, Fleming and Jessup [4]. This will be discussed in more detail in the next section.

Every commutative ABSL $\mathcal{L}$ on a complex separable Hilbert space H has the metric strong rank one density property. For, commutativity gives that the atoms of $\mathcal{L}$ are pairwise orthogonal. Thus the set of atoms is countable, say $\{L_i\}_1^\kappa$ (where possibly $\kappa = \infty$). Then, in the obvious way, Alg $\mathcal{L}$ is just the set of $\kappa \times \kappa$ diagonal matrices with operator entries, of the form diag (A_i) where $A_i \in \mathcal{B}(L_i)$ and $\sup_i \| A_i \| < \infty$. Since for each i the identity operator on L_i belongs to the strong operator closure of the unit ball of the set of finite rank operators of $\mathcal{B}(L_i)$ the result follows.

Most of our results concerning the transmission of the strong rank one density property from the old to the new also hold for the metric strong rank one density property with almost no change to their proofs. In particular, this is true if the new ABSL is obtained in either of the ways described in Example 2.7 (1) or by using Theorem 3.3. Also, 'strong rank one density' can be replaced by 'metric strong rank one density' in the statements of Theorems 4.2 and 4.3.

By virtue of Theorem 4.4 and the remarks contained in the preceding two paragraphs, the problem of which ABSL's on complex separable Hilbert spaces have the metric strong rank one density property reduces to consideration of ABSL's having property (G) . The following is a strengthening of Theorem 3.1, though for a more restricted class of spaces. (It is a type of Kaplansky density theorem but for non-self-adjoint algebras.)

THEOREM 4.6. *Every ABSL on a complex separable Hilbert space with precisely two atoms has the metric strong rank one density property.*

PROOF: Let $\mathcal{L}$ be an ABSL on a complex separable Hilbert space with precisely two atoms. If $\mathcal{L}$ is commutative the desired result follows almost immediately, and if $\mathcal{L}$ has property (G) it follows from

Theorem 4.5 . Suppose that $\mathcal{L}$ has neither of these properties. By Theorem 4.4, $\mathcal{L}$ is unitarily equivalent to a meshed product of ABSL's , $\mathcal{L}_1$ on H_1 and $\mathcal{L}_2$ on H_2 , where H_1 and H_2 are non-zero, complex, separable Hilbert spaces and where $\mathcal{L}_1$ is commutative and $\mathcal{L}_2$ has property (G) . Since $\mathcal{L}$ has precisely two atoms so does $\mathcal{L}_2$ ($\mathcal{L}_1$ has at most two atoms). By Theorem 4.5 $\mathcal{L}_2$ has the metric strong rank one density property, and by our earlier remarks so do $\mathcal{L}_1$ and $\mathcal{L}$. ∎

5. Strong M-bases

We begin this section by using the results of preceding sections to characterize those families $\{f_\gamma\}_\Gamma$ of vectors of a Banach space X for which $\{\langle f_\gamma \rangle\}_\Gamma$ is the set of atoms of an ABSL on X (Theorem 5.1). It turns out that these families are precisely those already known in the literature as strong M-bases.

The following terminology is standard and can be found, for example, in Singer [28]. A family $\{f_\gamma\}_\Gamma$ of vectors of a Banach space X is *complete* (in X) if $\vee_\Gamma f_\gamma = X$. A family $\{f_\gamma\}_\Gamma$ is *minimal* if

$$f_\gamma \notin \bigvee_{\mu \neq \gamma} f_\mu \text{ , for every } \gamma \in \Gamma .$$

Thus $\{f_\gamma\}_\Gamma$ is minimal if and only if there exists a family $\{f_\gamma^*\}_\Gamma \subseteq X^*$ biorthogonal to it. Note that if $\{f_\gamma\}_\Gamma$ is complete and minimal, the biorthogonal family $\{f_\gamma^*\}_\Gamma$ is uniquely determined. A family $\{f_\gamma^*\}_\Gamma \subseteq X^*$ is *total* (over X) if $\cap_\Gamma \ker f_\gamma^* = (0)$. A complete minimal family $\{f_\gamma\}_\Gamma$ with a total biorthogonal family $\{f_\gamma^*\}_\Gamma$ is called an *M-basis*. An M-basis $\{f_\gamma\}_\Gamma$ is called a *strong* M-*basis* if

$$x \in \vee\{f_\gamma : f_\gamma^*(x) \neq 0\} \text{ , for every } x \in X .$$

The notion of strong M-basis (or others equivalent to it) was introduced in the literature by several authors under various names. See for example Markus [21], Ruckle [27] and Singer [28]. (Actually the notion was introduced for sequences $\{f_n\}$, but as countability plays no role here, we drop this restriction.) In the terminology of Ruckle [27], for an M-basis, series summability implies finite series summability which in turn implies 1-series summability and the latter is equivalent to it being a strong

M-basis. Whether or not these three notions of summability actually coincide is an open question (see for example [28, Vol. II, p.270]), but it is known that strong series summability is a strictly stronger notion than all three [4]. As we shall see, the example of [4] shows that not every ABSL on a Banach space with the metric approximation property has the metric strong rank one density property. We improve the latter example somewhat (Example 5.5) by using the powerful constructions of Johnson [14] and Figiel and Johnson [7] concerning the failure of the bounded approximation property (these publications came after [4]) to construct a separable Banach space with the approximation property and with a series summable M-basis but which has no strongly series summable M-basis whatever. It is perhaps worth adding here that it is an open question whether or not every separable Banach space with the approximation property admits a series summable M-basis (see for example [28, Vol. II, p. 274]). Also, in [13] it is shown that every complex separable reflexive Banach space with the approximation property admits a strongly series summable M-basis.

The equivalence of the above definition of strong M-basis to other conditions is given in Theorem 5.1 below. Some of these equivalences are already known ([28, Vol. II]) but our interest here lies in the fact that strong M-bases arise precisely from ABSL's with one-dimensional atoms.

For comparison to Theorem 5.1 we mention the easily verifiable conditions that a complete and minimal family $\{f_\gamma\}_\Gamma$, with biorthogonal family $\{f^*_\gamma\}_\Gamma$, satisfies

(A1) $\cap_F \ker f^*_\gamma = \vee_{\Gamma\setminus F} f_\gamma$, for every finite subset $F \subseteq \Gamma$,

(A2) $\ker f^*_\gamma = \vee_{\mu\neq\gamma} f_\mu$, for every $\gamma \in \Gamma$,

(A3) $(\vee_F f_\gamma) \cap (\vee_{\Gamma\setminus F} f_\gamma) = (0)$, for every finite subset $F \subseteq \Gamma$.

Also we mention the following known equivalent conditions that a complete and minimal family $\{f_\gamma\}_\Gamma$ may or may not satisfy

(B1) $\{f_\gamma\}_\Gamma$ is an M-basis ,

(B2) $\cap_G \ker f_\gamma^* = \vee_{\Gamma\setminus G} f_\gamma$, for every finite or co-finite subset $G \subseteq \Gamma$,

(B3) $(\vee_I f_\gamma) \cap (\vee_J f_\gamma) = (0)$, for every pair I, J of disjoint subsets of Γ (there is no loss of generality in assuming that $J = \Gamma\setminus I$) ,

(B4) X is the approximate sum of $\{<f_\gamma>\}_\Gamma$.

An example of a complete minimal family not satisfying (B) is not hard to find (consider, for example, the family $\{f_n\}_1^\infty \subseteq \ell^2$ defined by $f_n = e_1 + e_{n+1}$; its biorthogonal family is $f_n^* = e_{n+1}$) . The following equivalences (C) imply (B) but not conversely. An example of a sequence satisfying (B) but not (C) is given by Markus [21], another example is in [28, Vol. II] and another in [30]. In the following theorem (for sequences) the equivalence of conditions (C2) and (C6) has been proved by Ruckle [27, Theorem 1.2] and the equivalence of (C2) and (C3) has been proved by Plans and Reyes (see [23]).

THEOREM 5.1. *Let* $\{f_\gamma\}_\Gamma$ *be a complete and minimal family of vectors of a Banach space* X , *with biorthogonal family* $\{f_\gamma^*\}_\Gamma \subseteq X^*$. *The following are equivalent.*

(C0) $\{f_\gamma\}_\Gamma$ *is a strong M-basis* ,

(C1) $\{<f_\gamma>\}_\Gamma$ *is the set of atoms of an* ABSL *on* X ,

(C2) $\cap_I \ker f_\gamma^* = \vee_{\Gamma\setminus I} f_\gamma$, *for every subset* $I \subseteq \Gamma$,

(C3) $(\vee_I f_\gamma) \cap (\vee_J f_\gamma) = \vee_{I \cap J} f_\gamma$, *for every pair* I, J *of subsets of* Γ ,

(C4) X *is the quasi-direct sum of* $\{<f_\gamma>\}_\Gamma$,

(C5) $\cap_\Lambda(\vee_{\Gamma_\lambda} f_\gamma) = \vee\{f_\gamma : \gamma \in \cap_\Lambda \Gamma_\lambda\}$,
for every family $\{\Gamma_\lambda\}_\Lambda$ *of subsets of* Γ ,

(C6) *For every* $\epsilon > 0$ *and every vector* $x \in X$ *there exists a finite sum* F *of the form* $F = \Sigma \lambda_\gamma(f_\gamma^* \otimes f_\gamma)$ (λ_γ *scalars*) *such that*

$$\| x - Fx \| < \epsilon .$$

PROOF: The equivalence of conditions (C1), (C3), (C4) and (C5) follows from Theorems 2.1 and 2.4. The equivalence of (C0) and (C2) follows almost immediately. We show that (C1) $\Rightarrow$ (C6) $\Rightarrow$ (C0) $\Rightarrow$ (C3) .

Assume (C1). Then (C6) follows from Theorem 2.5 and the fact that

$$(\vee_{\mu\neq\gamma} f_\mu)^\perp = <f_\gamma^*> , \text{ for every } \gamma \in \Gamma .$$

Assume (C6). Let $x \in X$ and put $I = \{\gamma \in \Gamma : f_\gamma^*(x) = 0\}$. Let $\epsilon > 0$ be arbitrary. By (C6) we have

$$\| x - \Sigma \lambda_\gamma(f_\gamma^* \otimes f_\gamma) x \| < \epsilon$$

for some finite sum $\Sigma \lambda_\gamma(f_\gamma^* \otimes f_\gamma)$. Clearly

$$\Sigma \lambda_\gamma(f_\gamma^* \otimes f_\gamma) x = \Sigma \lambda_\gamma f_\gamma^* (x) f_\gamma \in \vee_{\Gamma\setminus I} f_\gamma$$

since the terms that are zero can be discarded. Thus $x \in \vee\{f_\gamma : f_\gamma^*(x) \neq 0\}$ and (C0) holds.

Finally, assume (C0) and let I and J be subsets of Γ . Let $y \in (\vee_I f_\gamma) \cap (\vee_J f_\gamma)$. If $\gamma \notin I$ then $\vee_I f_\mu \subseteq \ker f_\gamma^*$ so $f_\gamma^*(y) = 0$. Similarly $f_\gamma^*(y) = 0$ if $\gamma \notin J$. Thus $\gamma \notin I \cap J$ implies $f_\gamma^*(y) = 0$ and so

$$y \in \vee\{f_\gamma : f_\gamma^*(y) \neq 0\} \subseteq \vee_{I \cap J} f_\gamma .$$

This shows that

$$(\vee_I f_\gamma) \cap (\vee_J f_\gamma) \subseteq \vee_{I \cap J} f_\gamma .$$

The reverse inclusion is obvious so (C3) holds. This completes the proof. ∎

In the case $\Gamma = \mathbb{Z}^+$ "for every pair I, J of subsets of Γ" in condition (C3) above can be replaced by "for every pair I, J of infinite subsets of $\mathbb{Z}^+$" [23] and by "for every pair I, J of infinite subsets of $\mathbb{Z}^+$ with infinite complements" [31] (see also [25]).

EXAMPLE 5.2.

(1) If $\{e_n\}_1^\infty$ is a basis of a Banach space X, then the natural projections $P_N = \sum_{i=1}^N e_i^* \otimes e_i$ converge to the identity operator in the strong operator topology (so the λ_γ's in (C6) above can all be taken to be unity). This is not always the case for strong M-basic sequences. For example, let $\{e_n\}_1^\infty$ be the usual orthonormal basis of ℓ^2 and define $\{f_n\}_1^\infty$ by $f_{2n-1} = e_{2n-1}$, $f_{2n} = e_{2n-1} + (1/2n)\, e_{2n}$ $(n \geq 1)$. Then $f_{2n-1}^* = e_{2n-1} - 2ne_{2n}$, $f_{2n}^* = 2ne_{2n}$ $(n \geq 1)$ is a biorthogonal sequence and it is readily checked, with $P_n = \sum_{i=1}^n f_i^* \otimes f_i$, that $P_{2n} = \sum_{i=1}^{2n} e_i \otimes e_i$ whereas

$$P_{2n-1} = \sum_{i=1}^{2n-1} e_i \otimes e_i - 2n\,(e_{2n} \otimes e_{2n-1}) .$$

If $x = \sum_1^\infty \frac{1}{n} e_n$, then

$$P_{2n-1}(x) = \sum_{i=1}^{2n-1} \frac{1}{i} e_i - e_{2n-1}$$

and $\sum_{i=1}^{2n-1} \frac{1}{i} e_i \to x$. Thus $\{P_{2n-1}x\}_1^\infty$ does not converge to x. Here the identity operator is the strong operator limit of a subsequence $(\{P_{2n}\}_1^\infty)$

of the sequence of natural projections but not of the sequence itself. A more elaborate example is the following.

(2) Let $\{f_n\}_1^\infty$ be the usual enumeration of the Fourier functions in $C[0,2\pi]$ the space of continuous functions with the uniform norm. Fejér's theorem states that every function $x \in C[0,2\pi]$ is the uniform limit of the arithmetic means of the partial sums of its Fourier series. That is, the λ_γ's in (C6) can be chosen (for sufficiently large n) as $\lambda_i = \frac{n+1-i}{n}$ and $Q_n = \sum_{i=1}^{n} \lambda_i(f_i^* \otimes f_i)$ converges strong operator to the identity operator. However it is not true that some subsequence $\{P_{n_k}\}_1^\infty$ of the sequence $\{P_n\}_1^\infty$ of natural projections converges strong operator to the identity. Indeed Menshov [22] (see also Bari [1, p.354]) has constructed a continuous function which is not the uniform limit of any subsequence of the sequence of partial sums of its Fourier series. (That is, the λ_γ's in (C6) cannot be replaced by ones.)

The authors thank Professor S. Pichorides for pointing out the reference to Menshov's example.

Below (see remarks following Theorem 5.8) we give another example where no subsequence of the sequence of natural projections converges strong operator to the identity operator, even though there is a net of natural projections so converging.

The following corollary extends Corollary 3.1 of Markus [21].

COROLLARY 5.3. *Let* X *be a reflexive Banach space. If the family of vectors* $\{f_\gamma\}_\Gamma$ *is a strong M-basis of* X *, then its biorthogonal family* $\{f_\gamma^*\}_\Gamma$ *is a strong M-basis of* X^* .

PROOF: By the preceding theorem, $\{f_\gamma\}_\Gamma$ is the set of atoms of an ABSL on X. Since

$$(\langle f_\gamma\rangle')^\perp = \langle f_\gamma^*\rangle , \text{ for every } \gamma \in \Gamma ,$$

Theorem 2.8 shows that $\{\langle f_\gamma^*\rangle\}_\Gamma$ is the set of atoms of an ABSL on X^*. The desired result now follows from Theorem 5.1. ∎

Clearly the condition of reflexivity cannot be omitted from the statement of the above corollary. A trivial example is any (Schauder) basis of ℓ^1. The span of its biorthogonal sequence cannot be dense in ℓ^∞ since the latter is not separable.

There is a dual version of the above corollary, coming 'down' from the dual. Here reflexivity, as expected, is not required.

COROLLARY 5.4. *If $\{f_\gamma\}_\Gamma$ is a family of vectors of a Banach space X and there is a strong M-basis of X^* which is biorthogonal to it, then $\{f_\gamma\}_\Gamma$ is a strong M-basis of X.*

PROOF: Let $\{f_\gamma^*\}_\Gamma$ be a strong M-basis of X^* biorthogonal to $\{f_\gamma\}_\Gamma$. The family biorthogonal to $\{f_\gamma^*\}_\Gamma$ is $\{\pi(f_\gamma)\}_\Gamma$ where $\pi : X \to X^{**}$ is the natural isometry. It is clear that $\{f_\gamma\}_\Gamma$ is complete and minimal. We establish condition (C2) of Theorem 5.1.

Let $I \subseteq \Gamma$ and let $x \in \cap_I \ker f_\gamma^*$. We show that $x \in \vee_{\Gamma\setminus I} f_\gamma$. Let $f^* \in X^*$ satisfy $f^*(\vee_{\Gamma\setminus I} f_\gamma) = \{0\}$ and let $\epsilon > 0$ be given. By (C6) applied to $\{f_\gamma^*\}_\Gamma$, there is a finite sum $\Sigma\, \lambda_\gamma(\pi(f_\gamma) \otimes f_\gamma^*)$ such that

$$\| f^* - (\Sigma\, \lambda_\gamma(\pi(f_\gamma) \otimes f_\gamma^*))\, f^* \| < \epsilon .$$

Thus

$$| f^*(x) - \Sigma\, \lambda_\gamma\, f^*(f_\gamma) f_\gamma^*(x) | \le \epsilon \| x \| .$$

But the sum inside the modulus signs is zero since, for $\gamma \in I$ we have $f_\gamma^*(x) = 0$ and for $\gamma \in \Gamma\backslash I$ we have $f^*(f_\gamma) = 0$. So $| f^*(x) | \leq \epsilon \| x \|$ and we deduce that $f^*(x) = 0$. An application of the Hahn-Banach theorem gives $x \in \vee_{\Gamma\backslash I} f_\gamma$ as required.

Thus

$$\cap_I \ker f_\gamma^* \subseteq \vee_{\Gamma\backslash I} f_\gamma$$

and since the reverse inclusion is obvious the proof is complete. ■

Notice that by using Theorem 2.10 instead of Theorem 2.8 in the proof of Corollary 5.3 we can recapture Lemma 3.1 of [21] which states that if $\{f_\gamma\}_\Gamma$ is a strong M-basis of a Hilbert space, then so is $\{f_\gamma\}_\Delta \cup \{f_\gamma^*\}_{\Gamma\backslash\Delta}$ for every subset $\Delta \subseteq \Gamma$. Several other corollaries to the results of the previous sections can be obtained by using the equivalence between strong M-bases and ABSL's with one-dimensional atoms proved in Theorem 5.1. In particular, the one-dimensional analogues of our results concerning the transmission of the (metric) strong rank one density property provide other corollaries.

As already mentioned Ruckle [27] calls strong M-bases 1-series summable M-bases. Strong M-bases whose corresponding ABSL (with one-dimensional atoms, see Theorem 5.1) has the strong rank one density property are precisely Ruckle's finitely series summable M-bases. Thus for example if $\{f_\gamma\}_\Gamma$ is a finitely series summable M-basis of a reflexive Banach space, then its biorthogonal family $\{f_\gamma^*\}_\Gamma$ is a finitely series summable M-basis of the dual space (this follows from Theorem 3.3); on a Hilbert space H, $\{f_\gamma\}_\Delta \cup \{f_\gamma^*\}_{\Gamma\backslash\Delta}$ is a finitely series summable M-basis of H, for every subset $\Delta \subseteq \Gamma$ (by Theorem 4.3).

It is an open question whether or not every 1-series summable M-basis is finitely series summable. In other words, it is an open question whether or not every ABSL with one-dimensional atoms has the strong rank one density property.†

Given a subspace lattice $\mathcal{L}$ on a Banach space X , the following question is stronger than the question of strong rank one density: Does there exist a uniformly bounded net of finite rank operators, each a finite sum of operators of Alg$\mathcal{L}$ of rank at most one, converging to the identity operator in the strong operator topology? Clearly (see [18,p.37]) if the answer is affirmative, then X has the bounded approximation property. However, even for an ABSL on a separable Banach space with the bounded approximation property the answer can be negative as we now explain. If X is a separable Banach space, the unit ball of $\mathcal{B}(X)$ is metrizable in the strong operator topology. Using this and the principle of uniform boundedness the question in the preceding paragraph, for ABSL's with one-dimensional atoms on a separable Banach space becomes: If $\{f_n\}_1^\infty$ is a strong M-basis is the identity operator a strong operator limit of a sequence of finite rank operators $\{F_n\}$ of the form

$$F_n = \sum_1^{N_n} \lambda_i^{(n)} (f_i^* \otimes f_i) .$$

In [27] strong M-bases for which the answer is affirmative are called strongly series summable. In [4] Crone, Fleming and Jessup give an example (Example 4.24) of a separable Banach space E with a basis $\{e_n\}_1^\infty$ in which there is a series summable strong M-basis $\{f_n\}_1^\infty$ which is not strongly series summable. Since E has a basis it has the bounded approximation property. For the ABSL with atoms $\{\langle f_n\rangle\}_1^\infty$, the answer to the question in the

---

† see ADDENDUM

preceding paragraph is negative. This example of [4] also shows that an ABSL with one-dimensional atoms on a Banach space with the metric approximation property need not have the metric strong rank one density property. For, there is an equivalent norm $|||\cdot|||$ on E such that $(E, |||\cdot|||)$ has the metric approximation property (indeed, if $\{P_n\}_1^\infty$ is the sequence of natural projections associated with the basis $\{e_n\}_1^\infty$, then $|||x||| = \sup_n \|P_n x\|$ defines such a norm [18,p.2]). In the Banach space $(E, |||\cdot|||)$ $\{f_n\}_1^\infty$ is still a strong M-basis which is not strongly series summable. Thus on $(E, |||\cdot|||)$ the ABSL with atoms $\{\langle f_n\rangle\}_1^\infty$ does not have the metric strong rank one density property.

The following improves the example of [4] somewhat. An M-basis $\{f_n\}_1^\infty$ of a Banach space X is *series summable* if and only if X has the approximation property and every nuclear map T on X satisfying $f_n^*(Tf_n) = 0$ $(n \in \mathbb{Z}^+)$ has trace zero [27, Theorem 1.2 D].

EXAMPLE 5.5.

There exists a separable Banach space X with the approximation property and with a series summable M-basis but which possesses no strongly series summable M-basis whatsoever.

We use the powerful constructions of [7,14]. To be specific, in [14] Johnson constructs a sequence $\{X_n\}_1^\infty$ of finite-dimensional Banach spaces such that the space

$$Y = (X_1 \oplus X_2 \oplus \ldots)_{\ell^1}$$

has a basis and the metric approximation property but whose dual space Y^* fails the approximation property. We proceed to define X following the example of [7, p.199]. By Theorem 1 of [7], for each $n \in \mathbb{Z}^+$, Y has an isomorphic copy Y_n failing the n-approximation property. Define X by

$$X = (Y_1 \oplus Y_2 \oplus \dots)_{\ell^2} .$$

This space has the approximation property but fails the bounded approximation property. In particular there is no sequence of finite rank operators on X converging to the identity operator in the strong operator topology. (A little more can be said: since X has the approximation property, there is no sequence of compact operators on X converging strong operator to the identity. This is so because if $\{K_n\}_1^\infty$ was such a sequence then choosing, by virtue of the approximation property, finite rank operators $\{F_n\}_1^\infty$ with $\| F_n - K_n \| \to 0$ we would get a contradiction.) Hence X admits no strongly series summable M-basis.

We now show that X has a series summable M-basis. For each $n \in \mathbb{Z}^+$ let $T_n : Y_n \to Y$ be an isomorphism. Let $\{e_j\}_1^\infty$ be a basis of Y. Consider the family $\{f_{ij} : i,j \in \mathbb{Z}^+\}$ of vectors of X given by

$$f_{ij} = (0,0,\dots,0,T_i^{-1} e_j,0,\dots)$$

where the non-zero entry is at the i-th coordinate. It can easily be verified that the family biorthogonal to $\{f_{ij}\}$ is $\{f_{ij}^* : i,j \in \mathbb{Z}^+\}$ where

$$f_{ij}^* = (0,0,\dots,0,T_i^* e_j^*,0,\dots)$$

where $\{e_j^*\}_1^\infty$ is the family biorthogonal to $\{e_j\}_1^\infty$. That $\{f_{ij}\}$ is an M-basis is obvious. To show that $\{f_{ij}\}$ is series summable, let $S = (S_{ij}) \in \mathcal{B}(X)$ be a nuclear map satisfying

$$f_{ij}^* (Sf_{ij}) = 0 , \text{ for every } i, j \in \mathbb{Z}^+ .$$

Then

$$0 = T_i^* e_j^* (S_{ii} T_i^{-1} e_j) = e_j^* (T_i S_{ii} T_i^{-1} e_j)$$

for all i and j so, as $\{e_j\}_1^\infty$ is a basis of Y, $\text{trace}(T_i S_{ii} T_i^{-1}) = 0$ for every i. Hence also trace $S_{ii} = 0$ for every i, and so trace $S = \Sigma_i \text{trace}(S_{ii}) = 0$. Thus $\{f_{ij}\}$ is series summable and we are finished.

Next we discuss a pointwise boundedness condition. In the following theorem, we use the fact that for an ABSL $\mathcal{L}$, Alg$\mathcal{L}$ is abelian if and only if every atom of $\mathcal{L}$ is one-dimensional [16].

THEOREM 5.6. *Let* $\{f_\gamma\}_\Gamma$ *be a strong M-basis of a Banach space* X *with biorthogonal family* $\{f_\gamma^*\}_\Gamma$. *If for every vector* $x \in X$ *there exists a constant* $K(x) > 0$ *such that for every* $\epsilon > 0$ *there exists a finite rank operator* F *of the form* $\Sigma\, \lambda_\gamma(f_\gamma^* \otimes f_\gamma)$ *(a finite sum) satisfying* $\| F \| \le K(x)$ and $\| x - Fx \| < \epsilon$, *then the identity operator on* X *belongs to the strong operator closure of the set of finite rank operators of the form* $\Sigma\, \lambda_\gamma(f_\gamma^* \otimes f_\gamma)$.

PROOF: We claim that there exists a constant $K_o > 0$ with the property that for every $\epsilon > 0$ and every vector $x \in X$ there exists a finite sum $F = \Sigma\, \lambda_\gamma(f_\gamma^* \otimes f_\gamma)$ such that $\| F \| \le K_o$ and $\| x - Fx \| < \epsilon$.

For $n \in \mathbb{Z}^+$ let

$$X_n = \{x \in X : \inf \| x - Fx \| = 0, \text{ the infimum being taken over all } F \text{ of the form } \Sigma\, \lambda_\gamma(f_\gamma^* \otimes f_\gamma) \text{ with } \| F \| \le n \}.$$

By hypothesis $\bigcup_{n=1}^\infty X_n = X$. Now each X_n is closed. For, let $\{x_k\}_1^\infty \subseteq X_n$ with $x_k \to x$ and let $\epsilon > 0$ be arbitrary. There exists $k \ge 1$ such that $\| x - x_k \| < \epsilon/2(n+1)$. For this x_k there exists $F = \Sigma\, \lambda_\gamma (f_\gamma^* \otimes f_\gamma)$ such that $\| F \| \le n$ and $\| x_k - Fx_k \| < \epsilon/2$. Then

$$\| x - Fx \| \le \| x - x_k \| + \| x_k - Fx_k \| + \| F \| \| x_k - x \| < \epsilon,$$

showing that $x \in X_n$. By the Baire category theorem there exist $x_o \in X$, $N \in \mathbb{Z}^+$ and $\delta > 0$ such that $\| y - x_o \| \le \delta$ implies $y \in X_N$.

Let $K_o = 2N + N^2$ and let $0 \neq x \in X$ and $\epsilon > 0$ be arbitrary. Clearly $y = \frac{\delta x}{\|x\|} + x_o$ satisfies $\| y - x_o \| \le \delta$ so there exist F and G each of the form $\Sigma \lambda_\gamma (f_\gamma^* \otimes f_\gamma)$ with $\| F \| \le N$ and $\| G \| \le N$ such that

$$\| (I - F) y \| < \frac{1}{2} \cdot \frac{\delta\epsilon}{N+1} \cdot \frac{1}{\|x\|} \quad \text{and} \quad \| (I - G)x_o \| < \frac{1}{2} \cdot \frac{\delta\epsilon}{N+1} \cdot \frac{1}{\|x\|} .$$

Since $I - F$ and $I - G$ commute we have

$$\begin{aligned} \| (I-F)(I-G) x \| &= \frac{\|x\|}{\delta} \| (I-F)(I-G)(y-x_o) \| \\ &\le \frac{\|x\|}{\delta} \| I-G \| \| (I-F)y \| + \frac{\|x\|}{\delta} \|I-F\| \|(I-G)x_o\| \\ &< \epsilon . \end{aligned}$$

Now $(I-F)(I-G) = I-E$ where $\|E\| = \| F + G - FG \| \le K_o$ and E has the form $\Sigma \lambda_\gamma (f_\gamma^* \otimes f_\gamma)$, completing the proof of the claim.

Using the above conclusion (and not using commutativity any more) we prove the theorem by the following argument.

Let vectors $x_1, x_2, \ldots, x_m \in X$ and $\epsilon > 0$ be given. Choose F_1, of the form $\Sigma \lambda_\gamma (f_\gamma^* \otimes f_\gamma)$, such that $\|F_1\| \le K_o$ and

$$\| (I-F_1)x_1 \| < \epsilon / (1 + K_o)^m .$$

If $F_1, F_2, \ldots, F_n$ $(1 \le n < m)$ have been chosen, chose for the inductive step F_{n+1} (of the form $\Sigma \lambda_\gamma (f_\gamma^* \otimes f_\gamma)$) such that $\| F_{n+1} \| \le K_o$ and (considering the vector $(I - F_n)(I - F_{n-1}) \ldots (I - F_1)x_{n+1}$)

$$\| (I - F_{n+1})(I - F_n)(I - F_{n-1}) \ldots (I - F_1)x_{n+1} \| < \epsilon/(1+K_o)^m .$$

Then, for $1 \le i \le m$ we have

$$\| (I - F_m)(I - F_{m-1}) \dots (I - F_1)x_i \|$$
$$\le \|(I-F_m)(I-F_{m-1}) \dots (I-F_{i+1})\| \cdot \|(I-F_i)(I-F_{i-1}) \dots (I-F_1)x_i \|$$
$$\le (1 + K_o)^{m-i} \frac{\epsilon}{(1+K_o)^m} < \epsilon .$$

Thus if $F = I - (I-F_m)(I-F_{m-1}) \dots (I-F_1)$ then F is of the form $\Sigma \lambda_\gamma (f_\gamma^* \otimes f_\gamma)$ and

$$\|x_i - Fx_i \| < \epsilon , \text{ for } i = 1,2,\dots, m .$$

This completes the proof. ■

We remark that the F's obtained in the last part of the above proof are not necessarily uniformly bounded (independently of $x_1, x_2, \dots, x_m$ and ϵ) ; the argument merely gives $\| F \| \le 1 + (1 + K_o)^m$.

We have already mentioned in this section several methods by which new strong M-bases can be obtained from old ones. Also, every (Schauder) basis is a strong M-basis, so is every permutation of a strong M-basis; so too is every image of a strong M-basis under a bicontinuous bijection.

Below, for every $1 \le p < \infty$, we give examples of sequences $\{f_n\}_1^\infty$ each of which is a strong M-basis of ℓ^p . Some $\{f_n\}$ are bases and some are not, but even in the case of ℓ^2 , this class of examples is as unlike an orthonormal basis as anything can be. For each such sequence, the corresponding ABSL (with atoms $\{<f_n>\}_1^\infty$) has the strong rank one density property. In fact, in each case there is a sequence $\{F_n\}_1^\infty$ of finite rank operators of the form $\Sigma \lambda_i (f_i^* \otimes f_i)$ converging strong operator to the identity operator (so each $\{f_n\}_1^\infty$ is strongly series summable). Also, there is always a net of natural projections $P_N = \sum_1^N f_j^* \otimes f_j$ converging strong operator to the identity operator, but not necessarily a sequence of natural projections with this property.

Let $\{e_n\}_1^\infty$ be the usual basis for ℓ^p, $1 \le p < \infty$. If $\{a_n\}_1^\infty$ is a sequence of scalars put

$$f_n = (a_1, a_2, \ldots, a_n, 0, 0, \ldots), \text{ for } n \ge 1 .$$

It is easily proved that $\{f_n\}_1^\infty$ is complete in ℓ^p if and only if $a_n \ne 0$ for every $n \ge 1$. Let this be the case. The sequence $f_n^* \subseteq (\ell^p)^*$ defined by $f_n^* = \frac{e_n}{a_n} - \frac{e_{n+1}}{a_{n+1}}$ is biorthogonal to $\{f_n\}_1^\infty$ so $\{f_n\}_1^\infty$ is minimal. For $N \in \mathbb{Z}^+$ let P_N be the natural projection operator defined by $P_N = \sum_1^N f_n^* \otimes f_n$. Since

$$\bigcap_1^\infty \ker f_n^* = \{\{\xi_i\}_1^\infty \in \ell^p : \text{for some scalar } \lambda,\ \xi_i = \lambda a_i \text{ for every } i \in \mathbb{Z}^+\},$$

$\{f_n^*\}_1^\infty$ is total over ℓ^p if and only if $\{a_n\}_1^\infty \notin \ell^p$. Even more is true in this case.

THEOREM 5.7. *The sequence* $\{f_n\}_1^\infty$ *(defined as above) is a strong* M-*basis of* ℓ^p $(1 \le p < \infty)$ *if and only if* $\{a_n\}_1^\infty \notin \ell^p$. *In this case, for every* $\epsilon > 0$ *and every finite set* $x_1, x_2, \ldots, x_m$ *of vectors of* ℓ^p, *there exists a natural projection* P_N *such that* $\| x_i - P_N x_i \| < \epsilon$, *for* $i = 1, 2, \ldots, m$.

PROOF: If $\{f_n\}_1^\infty$ is a strong M-basis of ℓ^p, then $\{f_n^*\}_1^\infty$ is total over ℓ^p by definition, so $\{a_n\}_1^\infty \notin \ell^p$.

Conversely, suppose that $\{a_n\}_1^\infty \notin \ell^p$. Then

$$\sum_{n=1}^{\infty} \frac{|a_{n+1}|^p}{|a_1|^p + |a_2|^p + \ldots + |a_n|^p}$$

diverges. For if S_k denotes the k-th partial sum of this series then, for $k > \ell \geq 1$, we have

$$| S_k - S_\ell| = \frac{|a_{\ell+2}|^p}{|a_1|^p + \ldots + |a_{\ell+1}|^p} + \ldots + \frac{|a_{k+1}|^p}{|a_1|^p + \ldots + |a_k|^p}$$

$$\geq 1 - \frac{|a_1|^p + \ldots + |a_{\ell+1}|^p}{|a_1|^p + \ldots + |a_{k+1}|^p} \to 1 \quad \text{as} \quad k \to \infty ,$$

so the Cauchy criterion fails.

Let $\epsilon > 0$ be arbitrary and let $x = \{\xi_j\}_1^\infty \in \ell^p$. We show that there is a natural projection P_N such that $\| x - P_N x \| < \epsilon$. (The general case of m vectors $x_1, x_2, \ldots, x_m$ will follow from this case.) Routine calculation shows that

$$\| x - P_n x \|^p = \left(\frac{|a_1|^p + \ldots + |a_n|^p}{|a_{n+1}|^p} \right) |\xi_{n+1}|^p + \sum_{n+1}^{\infty} |\xi_j|^p . \tag{1}$$

Now there exists $M \in \mathbb{Z}^+$ such that $\sum_{n+1}^{\infty} |\xi_j|^p < \epsilon^p/2$, for every $n \geq M$. For this M we claim that there exists an $N \geq M$ such that

$$\left(\frac{|a_1|^p + \dots + |a_N|^p}{|a_{N+1}|^p}\right) |\xi_{N+1}|^p < \epsilon^p/2 .$$

Indeed, otherwise we would have

$$\frac{|a_{n+1}|^p}{|a_1|^p + \dots + |a_n|^p} \le \frac{2}{\epsilon^p} |\xi_{n+1}|^p ,$$

for every $n \ge M$. But this contradicts the fact that

$$\sum_{n=1}^{\infty} \frac{|a_{n+1}|^p}{|a_1|^p + \dots + |a_n|^p}$$

diverges. Thus $\| x - P_N x \| < \epsilon$ for some $N \ge M$. By Theorem 5.1, (C6), $\{f_n\}_1^\infty$ is a strong M-basis.

Now let $x_1, x_2, \dots, x_m \in \ell^p$ be given. Let $x_i = \{\xi_{ij}\}_{j=1}^\infty$. For each $i = 1,2,\dots,m$ the sequence $|x_i| = \{|\xi_{ij}|\}_{j=1}^\infty$ belongs to ℓ^p and so does the sum

$$x = \sum_{i=1}^m |x_i| = \{ \sum_{i=1}^m | \xi_{ij} | \}_{j=1}^\infty .$$

The j-th coordinate $\sum_{i=1}^m |\xi_{ij}|$ of x dominates the j-th coordinate $|\xi_{ij}|$ of $|x_i|$ so (1) gives

$$\|x - P_n x \|^p \ge \|x_i - P_n x_i\|^p , \text{ for every } i = 1,2,\dots,m,$$

and for every $n \in \mathbb{Z}^+$. If we choose $N \in \mathbb{Z}^+$ such that $\|x - P_N x \| < \epsilon$ (see the first part of the proof), then

$$\|x_i - P_N x_i\| < \epsilon , \text{ for every } i = 1,2,\dots,m .$$

This completes the proof. ■

REMARKS

(1) In the statement of the above theorem we can replace ℓ^p with c_o. The proof is similar. For example, we need the observations

(i) $\{f_n^*\}_1^\infty$ is total over c_o if and only if $\{a_n\}_1^\infty \notin c_o$,

(ii) If $\{a_n\}_1^\infty \notin c_o$ then

$$\left\{ \frac{|a_{n+1}|}{\max\limits_{1\le i\le n} |a_i|} \right\}_1^\infty \notin c_o ,$$

and (iii) equation (1) becomes

$$\|x - P_n x\|_\infty = \sup \left(\left\{ \frac{|a_i| \, |\xi_{n+1}|}{|a_{n+1}|} \right\}_{i=1}^\infty \cup \{|\xi_j|\}_{j=n+1}^\infty \right)$$

$$\le \left(\frac{\max\limits_{1\le i\le n} |a_i|}{|a_{n+1}|} \right) |\xi_{n+1}| + \sup_{j\ge n+1} |\xi_j| .$$

(2) Also let us mention that, in ℓ^2, this class of strong M-bases has no member similar to an orthonormal basis. For, if $\{Sf_n\}_1^\infty$ were an orthonormal basis for some invertible operator S, then letting U be the unique unitary operator satisfying $Ue_n = Sf_n$ $(n \in \mathbb{Z}^+)$, we would have $S^{-1}Ue_n = f_n = \sum_{i=1}^n a_i e_i$. Thus the matrix of $S^{-1}U$ relative to $\{e_n\}_1^\infty$ would have constant first row (each term equals $a_1 \neq 0$) and so $S^{-1}U$ is unbounded. Contradiction.

(3) In conjunction to Corollary 5.3, in the case of ℓ^1 (with $\{a_n\}_1^\infty \notin \ell^1$) the biorthogonal family $\{f_n^*\}_1^\infty$ is clearly not complete in

$(\ell^1)^* = \ell^\infty$. We show that in fact $\overset{\infty}{\underset{1}{\vee}} f_n^* = c_o$. The inclusion $\overset{\infty}{\underset{1}{\vee}} f_n^* \subseteq c_o$ is obvious since the coordinates of each f_n^* are eventually zero. Conversely, if $n \in \mathbb{Z}^+$ and

$$\lambda_i = a_1 \frac{|a_{i+1}| + \ldots + |a_{n+1}|}{|a_1| + \ldots + |a_{n+1}|}, \text{ for } i = 1,2,\ldots,n,$$

then it is easy to see that

$$\| \lambda_1 f_1^* + \ldots + \lambda_n f_n^* - e_1 \|_\infty = \frac{|a_1|}{|a_1| + \ldots + |a_n|} \to 0,$$

so that $e_1 \in \overset{\infty}{\underset{1}{\vee}} f_n^*$. For $n \geq 2$,

$$e_n = a_n \frac{e_1}{a_1} - a_n \left(\sum_{i=1}^{n-1} f_i^* \right),$$

so it follows that $e_n \in \overset{\infty}{\underset{1}{\vee}} f_n^*$. The desired result easily follows.

Let us continue our investigation of the properties of the sequence $\{f_n\}_1^\infty$ with $\{a_n\}_1^\infty \notin \ell^p$ considered above. This family $\{f_n\}_1^\infty$ may or may not be a basis. The relevant criterion is given below.

THEOREM 5.8. *The sequence* $\{f_n\}_1^\infty$ *considered in Theorem* 5.7 *is a basis of* ℓ^p $(1 \leq p < \infty)$ *if and only if*

$$\sup_n \frac{|a_1|^p + \ldots + |a_n|^p}{|a_{n+1}|^p} < \infty .$$

In particular, if $\liminf \left| \frac{a_{n+1}}{a_n} \right| > 1$, *then* $\{f_n\}_1^\infty$ *is a basis, and if* $\{f_n\}_1^\infty$ *is a basis, then* $\limsup \left| \frac{a_{n+1}}{a_n} \right| > 1$.

PROOF: As is well known, $\{f_n\}_1^\infty$ is a basis if and only if $\{\|P_n\|\}_1^\infty$ is bounded, where P_n is the n-th natural projection. The first part of the theorem will be proved if we show that, for every $n \in \mathbb{Z}^+$

$$\left(\frac{|a_1|^p + \ldots + |a_n|^p}{|a_{n+1}|^p} \right)^{1/p} \leq \| P_n \| \leq 1 + \left(\frac{|a_1|^p + \ldots + |a_n|^p}{|a_{n+1}|^p} \right)^{1/p} .$$

It is easily verified that $P_n = E_n - \frac{1}{a_{n+1}} (e_{n+1} \otimes f_n)$

where E_n is the norm one projection onto $\bigvee_1^n e_j$ along $\bigvee_{n+1}^\infty e_j$. Using this, the right hand inequality above follows from the triangle inequality. The left hand inequality is simply the statement that $\|P_n e_{n+1}\| \leq \|P_n\|$ since

$$\|P_n e_{n+1}\| = \|E_n e_{n+1} - \frac{1}{a_{n+1}} f_n\|$$

$$= \frac{1}{|a_{n+1}|} \|f_n\| .$$

Now suppose that $\liminf \left| \frac{a_{n+1}}{a_n} \right| > 1$. Let r satisfy $\liminf \left| \frac{a_{n+1}}{a_n} \right| > r > 1$. Then, by the definition of limit inferior, there is a positive integer N such that for every $n \geq N$ we have $\left| \frac{a_{n+1}}{a_n} \right| > r$.

So for $n > N$ we have

$$\frac{|a_1|^p + \dots + |a_n|^p}{|a_{n+1}|^p} = \frac{|a_1|^p + \dots + |a_N|^p}{|a_{n+1}|^p} + \frac{|a_{N+1}|^p + \dots + |a_n|^p}{|a_{n+1}|^p}$$

$$\leq \frac{|a_1|^p + \dots + |a_N|^p}{|a_N|^p} + \left(\frac{1}{r^p} + \frac{1}{r^{2p}} + \dots + \frac{1}{r^{(n-N)p}} \right)$$

which is uniformly bounded (the geometric progression is bounded by $(r^p - 1)^{-1}$) . Hence $\{f_n\}_1^\infty$ is a basis.

Conversely, suppose that $\{f_n\}_1^\infty$ is a basis. As $\{a_n\}_1^\infty \notin \ell^p$ we have $\limsup \left| \frac{a_{n+1}}{a_n} \right| \geq 1$ by the ratio test. We show that we cannot have $\limsup \left| \frac{a_{n+1}}{a_n} \right| = 1$. On the contrary if this were the case, then for any fixed $\epsilon > 0$ there would be a positive integer N such that $|a_{n+1}| < \sqrt[p]{1+\epsilon}\, |a_n|$, for every $n \geq N$. Then, for $n > N$ we would have

$$\frac{|a_1|^p + \dots + |a_n|^p}{|a_{n+1}|^p} \geq \frac{|a_N|^p + \dots + |a_n|^p}{|a_{n+1}|^p}$$

$$\geq \sum_{j=N}^{n} \left(\frac{1}{1+\epsilon} \right)^{j-N+1}$$

$$= \frac{1}{\epsilon} \left[1 - \left(\frac{1}{1+\epsilon} \right)^{n+1-N} \right] .$$

Since

$$\frac{1}{\epsilon}\left[1-\left(\frac{1}{1+\epsilon}\right)^{n+1-N}\right] \to \frac{1}{\epsilon}, \text{ as } n \to \infty ,$$

it would follow that, for all sufficiently large n

$$\frac{|a_1|^p + \dots + |a_n|^p}{|a_{n+1}|^p} \geq \frac{1}{2\epsilon} .$$

This contradicts the fact that

$$\sup_n \frac{|a_1|^p + \dots + |a_n|^p}{|a_{n+1}|^p} < \infty . \blacksquare$$

The two implications of the preceding theorem are not reversible. For example, if $a_{2n-1} = 1$ and $a_{2n} = 2$ $(n \in \mathbb{Z}^+)$, then $\limsup \left|\frac{a_{n+1}}{a_n}\right| > 1$, yet the corresponding sequence $\{f_n\}_1^\infty$ is not a basis (by the first part of the theorem). Also $a_{2n-1} = a_{2n} = 2^n$ gives a basis, but $\liminf \left|\frac{a_{n+1}}{a_n}\right| = 1$. Of course, if $\lim_n \left|\frac{a_{n+1}}{a_n}\right|$ exists, then the theorem shows that $\{f_n\}_1^\infty$ is a basis if and only if this limit is strictly greater than one.

In the case of C_o, a similar proof shows that $\{f_n\}_1^\infty$ is a basis if and only if $\left\{\max_{1\leq i\leq n} |a_i| / |a_{n+1}|\right\}_1^\infty$ is bounded. This for example shows (the well known fact) that $f_n = \sum_{i=1}^n e_i$ is a basis of C_o but not of ℓ^p $(1 \leq p < \infty)$. It is however a strong M-basis of ℓ^p $(1 \leq p < \infty)$ by Theorem 5.7. On the other hand, $f_n = \sum_{i=1}^n r^{i-1} e_i$, where $r > 1$, is a basis of both ℓ^p and C_o.

In Theorem 5.7 we showed that, for the strong M-basis $\{f_n\}_1^\infty$ considered, there is a net of natural projections converging to the identity operator in the strong operator topology. Since we can choose $\{a_n\}_1^\infty \notin \ell^p$ so that $\{\|P_N\|\}_1^\infty$ has no bounded subsequences (See the proof of Theorem 5.8. For example we can take $a_n = 1$ for every n.) the uniform boundedness principle shows that it is possible to arrange that no subsequence of $\{P_N\}_1^\infty$ converges strong operator to the identity.

However, in the next theorem we exhibit a sequence $\{F_n\}_1^\infty$ of finite rank operators of $\mathrm{Alg}\{\langle f_n\rangle : n \in \mathbb{Z}^+\}$ converging to the identity operator in the strong operator topology.

THEOREM 5.9. *Let* $\{f_n\}_1^\infty$ *be the strong* M-*basis of* ℓ^p $(1 \le p < \infty)$ *as considered in Theorem* 5.7. *The sequence* $\{F_n\}_1^\infty$ *of finite rank operators given by*

$$F_n = \sum_{j=1}^{n} \left(\frac{|a_{j+1}|^p}{|a_1|^p + \ldots + |a_{n+1}|^p} \right) P_j ,$$

where P_j *is the* j-*th natural projection*, $P_j = \sum_{i=1}^{j} f_i^* \otimes f_i$, *converges to the identity operator in the strong operator topology. Consequently,* $\{f_n\}_1^\infty$ *is a strongly series summable* M-*basis of* ℓ^p .

PROOF: For $1 \le i \le n$ we have

$$F_n f_i = \left(1 - \frac{|a_1|^p + \ldots + |a_i|^p}{|a_1|^p + \ldots + |a_{n+1}|^p} \right) f_i .$$

Thus, since $\{a_n\}_1^\infty \notin \ell^p$, $F_n f_i \to f_i$ as $n \to \infty$ for each fixed $i \in \mathbb{Z}^+$. It follows that $F_n x \to x$ for every x belonging to the linear span of $\{f_i\}_1^\infty$, which is dense in ℓ^p . The proof is completed by showing that

$\{\|F_n\|\}_1^\infty$ is bounded. With E_j as in the proof of the preceding theorem, so that

$$P_j = E_j - \frac{1}{a_{j+1}} (e_{j+1} \otimes f_j) ,$$

we have $F_n = D_n - B_n$ where

$$D_n = \frac{1}{|a_1|^p + \dots + |a_{n+1}|^p} \sum_{j=1}^{n} |a_{j+1}|^p E_j$$

and

$$B_n = \frac{1}{|a_1|^p + \dots + |a_{n+1}|^p} \sum_{j=1}^{n} \frac{|a_{j+1}|^p}{a_{j+1}} (e_{j+1} \otimes f_j) .$$

Now, the matrix of D_n relative to $\{e_n\}_1^\infty$ is diagonal with finitely many non-zero entries, all positive, so $\|D_n\|$ is the largest diagonal entry. This gives

$$\|D_n\| = \frac{1}{|a_1|^p + \dots + |a_{n+1}|^p} \sum_{j=1}^{n} |a_{j+1}|^p \leq 1 .$$

Let $1 < p < \infty$ and define q by $\frac{1}{p} + \frac{1}{q} = 1$. If $x = \{\xi_i\}_1^\infty \in \ell^p$ we use Hölder's inequality to obtain

$$\| \sum_{j=1}^{n} \frac{|a_{j+1}|^p}{a_{j+1}} (e_{j+1} \otimes f_j) x \|^p = \sum_{k=1}^{n} |a_k|^p \left| \sum_{j=k}^{n} \frac{|a_{j+1}|^p}{a_{j+1}} \xi_{j+1} \right|^p$$

$$\leq \sum_{k=1}^{n} |a_k|^p \left(\sum_{j=k}^{n} |a_{j+1}|^p \right)^{p/q} \left(\sum_{i=k}^{n} |\xi_{i+1}|^p \right)$$

$$\leq \sum_{k=1}^{n+1} |a_k|^p \left(\sum_{j=1}^{n+1} |a_j|^p \right)^{p/q} \|x\|^p$$

$$= \left(\sum_{k=1}^{n+1} |a_k|^p \right)^p \|x\|^p .$$

Thus $\|B_n x\|^p \leq \|x\|^p$, so $\|B_n\| \leq 1$.

A similar argument shows that $\|B_n x\| \leq \|x\|$ even in the case $p = 1$; the Hölder inequality being replaced by its '$q = \infty$' version, the triangle inequality here:

$$\left| \sum_{j=k}^{n} \frac{|a_{j+1}|}{a_{j+1}} \xi_{j+1} \right| \leq \sum_{j=k}^{n} |\xi_{j+1}| .$$

Therefore we have, for every $1 \leq p < \infty$, $\|F_n\| \leq \|D_n\| + \|B_n\| \leq 2$ and the proof is complete. ∎

Note that the case $a_n = 1$ $(n \in \mathbb{Z}^+)$ and $p = 2$ was considered in [6], where it was shown that the sequence of Cesàro means $\frac{1}{n}(P_1 + P_2 + \ldots + P_n)$ converges to the identity operator in the strong operator topology.

We close this section with a corollary to Theorem 5.7.

COROLLARY 5.10. *Let* $\{x_n\}_1^\infty$ *be a sequence of non-zero vectors in a Banach space* X . *Let* $1 \leq p < \infty$ *and for each* $n \in \mathbb{Z}^+$ *let* f_n *be the element of* $\ell^p(X)$ *defined by* $f_n = (x_1, x_2, \ldots, x_n, 0, 0, \ldots)$. *The sequence* $\{f_n\}_1^\infty$ *is a strong* M*-basis of* $\bigvee_1^\infty f_n$ *if and only if* $\{\|x_n\|\}_1^\infty \notin \ell^p$.

PROOF: It is easily verified that

$$\bigvee_1^\infty f_n = \{\{y_j\}_1^\infty \in \ell^p(X) : y_j \in \langle x_j \rangle \text{ , for every } j \in \mathbb{Z}^+\} .$$

The mapping $S : \bigvee_1^\infty f_n \to \ell^p$ defined by

$$S(\lambda_1 x_1, \lambda_2 x_2, \ldots) = (\lambda_1 \|x_1\|, \lambda_2 \|x_2\|, \ldots)$$

is a surjective linear isometry. Thus $\{f_n\}_1^\infty$ is a strong M-basis of $\bigvee_1^\infty f_n$ if and only if $\{Sf_n\}_1^\infty$ is a strong M-basis of ℓ^p. Since $Sf_n = (\|x_1\|, \|x_2\|, \ldots, \|x_n\|, 0, 0, \ldots)$ for $n \in \mathbb{Z}^+$, the result follows from Theorem 5.7. ∎

6. Selecting and slicing

Methods of obtaining new ABSL's from old ones have been described in the previous sections. Here we consider another possible method. Does 'slicing' the atoms of an ABSL produce another? Precisely, if $\{L_\gamma\}_\Gamma$ is the set of atoms of an ABSL on X and if, for every $\gamma \in \Gamma$, K_γ is a non-zero subspace of L_γ, does it follow that $\{K_\gamma\}_\Gamma$ is the set of atoms of an ABSL on $\vee_\Gamma K_\gamma$? Perhaps surprisingly, the answer is negative even for the case of three atoms in Hilbert space (the answer is trivially positive for two atoms); an example is given below (Theorem 6.4). Observe that, in general, $\vee_\Gamma K_\gamma$ is always the approximate sum of the 'slices' $\{K_\gamma\}_\Gamma$ since for every pair I, J of disjoint subsets of Γ we have

$$(0) \subseteq (\vee_I K_\gamma) \cap (\vee_J K_\gamma) \subseteq (\vee_I L_\gamma) \cap (\vee_J L_\gamma) = (0) \quad .$$

So our example provides another example, different from that given in [3], of an approximate sum of three subspaces that is not quasi-direct. It has the added property that it 'sits' in a quasi-direct sum.

Before giving the example we prove some positive results.

THEOREM 6.1. *Let* $\{L_\gamma\}_\Gamma$ *be the set of atoms of an* ABSL *on a Banach space* X . *If* Δ *is a finite subset of* Γ *and* $\{K_\gamma\}_\Delta$ *is a family of non-zero subspaces of* X *satisfying*

(i) $K_\gamma \subseteq L_\gamma$, *for every* $\gamma \in \Delta$,

and (ii) $K_\gamma + \bigvee_{\mu\neq\gamma} L_\mu$ *is closed, for every* $\gamma \in \Delta$,

then $\{K_\gamma\}_\Delta \cup \{L_\gamma\}_{\Gamma\setminus\Delta}$ *is the set of atoms of an* ABSL *on* $(\vee_\Delta K_\gamma) \vee (\vee_{\Gamma\setminus\Delta} L_\gamma)$.

PROOF: If M and N are subspaces of X satisfying $M \cap N = (0)$ with $M + N$ closed, then $M_o + N_o$ is also closed, for all subspaces M_o and N_o satisfying $M_o \subseteq M$, $N_o \subseteq N$. This shows that it is enough to prove the result for $\Delta = \{\lambda\}$, a singleton.

By Theorem 2.4 we have to show that

$$(\vee_I K_\gamma) \cap (\vee_J K_\gamma) = \vee_{I \cap J} K_\gamma ,$$

for every pair I, J of subsets of Γ, where we have taken $K_\gamma = L_\gamma$ if $\gamma \neq \lambda$. If $\lambda \notin I \cap J$ this is certainly true since then

$$(\vee_I K_\gamma) \cap (\vee_J K_\gamma) \subseteq (\vee_I L_\gamma) \cap (\vee_J L_\gamma) = \vee_{I \cap J} L_\gamma = \vee_{I \cap J} K_\gamma$$

and the reverse inclusion is obvious.

Now suppose that $\lambda \in I \cap J$. We only have to show that

$$(K_\lambda + \vee_{I_o} L_\gamma) \cap (K_\lambda + \vee_{J_o} L_\gamma) \subseteq K_\lambda + \vee_{I_o \cap J_o} L_\gamma \qquad (*)$$

where $I_o = I \setminus \{\lambda\}$ and $J_o = J \setminus \{\lambda\}$. Let the vector $x \in X$ belong to the left hand side of (*). Then $x \in K_\lambda + \vee_{\gamma \neq \lambda} L_\gamma$ so $x - x_\lambda \in \vee_{\gamma \neq \lambda} L_\gamma$ for some vector $x_\lambda \in K_\lambda$. As

$$x \in K_\lambda + \vee_{I_o} L_\gamma \subseteq \vee_I L_\gamma$$

we also have $x - x_\lambda \in \vee_I L_\gamma$. Similarly $x - x_\lambda \in \vee_J L_\gamma$ and so

$$x - x_\lambda \in (\vee_{\gamma \neq \lambda} L_\gamma) \cap (\vee_I L_\gamma) \cap (\vee_J L_\gamma) = \vee_{I_o \cap J_o} L_\gamma .$$

Hence $x \in K_\lambda + \vee_{I_o \cap J_o} L_\gamma$, as required. ■

In [30, Proposition II] it is shown that there exists a strong M-basis $\{f_i\}_1^\infty$ of a Banach space X and an increasing sequence $\{n_i\}_1^\infty$ of integers and there exist non-zero vectors

$$g_i \in \vee \{f_j : n_{i-1}+1 \leq j \leq n_i\} \text{ (where } n_o = 0)$$

such that $\{g_i\}_1^\infty$ fails to be a strong M-basis of $\overset{\infty}{\underset{1}{\vee}} g_i$.

Using the results of the preceding section the subspaces $\{<f_i>\}_1^\infty$ form the set of atoms of an ABSL on X and hence so do $\{L_i\}_1^\infty$ where

$$L_i = \vee\{f_j : n_{i-1} + 1 \leq j \leq n_i\} .$$

Taking $K_i = <g_i> \ (i \in \mathbb{Z}^+)$, it follows that the constraint in the statement of Theorem 6.1 that Δ be finite cannot be dropped. Notice also that $\overset{\infty}{\underset{1}{\vee}} g_i$ is the approximate, but not the quasi-direct, sum of $\{<g_i>\}_1^\infty$.

Related to this we prove the following.

THEOREM 6.2. *Let* $\{L_\gamma\}_\Gamma$ *be the set of atoms of an* ABSL *on a Banach space* X . *For every* $\gamma \in \Gamma$, *let* $0 \neq f_\gamma \in L_\gamma$. *Then* $\{<f_\gamma>\}_\Gamma$ *is the set of atoms of an* ABSL *on* $K = \vee_\Gamma f_\gamma$ *if and only if*

$$(\vee_I L_\gamma) \cap K = \vee_I f_\gamma \text{ , for every subset } I \subseteq \Gamma .$$

PROOF: First we show that the stated condition is sufficient. Suppose it holds. Since $\{f_\gamma\}_\Gamma$ is obviously complete and minimal in K , by Theorem 5.1 it is enough to show that

$$(\vee_I f_\gamma) \cap (\vee_J f_\gamma) = \vee_{I \cap J} f_\gamma ,$$

for every pair I, J of subsets of Γ . Using Theorem 2.4 we have

$$\begin{aligned}(\vee_I f_\gamma) \cap (\vee_J f_\gamma) &\subseteq (\vee_I L_\gamma) \cap (\vee_J L_\gamma) \cap K \\ &= (\vee_{I \cap J} L_\gamma) \cap K \\ &= \vee_{I \cap J} f_\gamma ,\end{aligned}$$

and the reverse inclusion is obvious.

Conversely, suppose that $\{<f_\gamma>\}_\Gamma$ is the set of atoms of an ABSL on K. For each $\gamma \in \Gamma$, $f_\gamma \notin L'_\gamma$ so, by the Hahn–Banach theorem, $g^*_\gamma(f_\gamma) = 1$ for some $g^*_\gamma \in (L'_\gamma)^\perp$. Let $f^*_\gamma = g^*_\gamma\big|_K$. Then since $f_\mu \in L'_\gamma$ for $\mu \neq \gamma$ (because we have $L'_\gamma = \vee_{\mu\neq\gamma} L_\mu$) it follows that in fact $f^*_\gamma(f_\mu) = g^*_\gamma(f_\mu) = \delta_{\gamma\mu}$. Thus $\{f^*_\gamma\}_\Gamma \subseteq K^*$ is biorthogonal to $\{f_\gamma\}_\Gamma$. Let I be any subset of Γ and let $x \in (\vee_I L_\gamma) \cap K$. If $\gamma \notin I$ then $f^*_\gamma(x) = 0$ since

$$\vee_I L_\mu \subseteq L'_\gamma \subseteq \ker g^*_\gamma .$$

But by Theorem 5.1, $\{f_\gamma\}_\Gamma$ is a strong M-basis of K, so $x \in \vee\{f_\gamma : f^*_\gamma(x) \neq 0\}$. Thus $x \in \vee_I f_\gamma$. This shows that

$$(\vee_I L_\gamma) \cap K \subseteq \vee_I f_\gamma .$$

The reverse inclusion is obvious and the proof is complete. ∎

With the hypotheses of the above theorem, it is not hard to show that $(\vee_I L_\gamma) \cap K = \vee_I f_\gamma$ if I is a subset of Γ for which $(\vee_I f_\gamma) + (\vee_{\Gamma\setminus I} f_\gamma)$ is closed (for example if I or $\Gamma\setminus I$ is finite). Thus if $(\vee_I L_\gamma) + (\vee_J L_\gamma)$ is closed for every pair I, J of subsets of Γ, then $\{<f_\gamma>\}_\Gamma$ is the set of atoms of an ABSL on $\vee_\Gamma f_\gamma$, for any 'selection' $\{f_\gamma\}_\Gamma$ satisfying $0 \neq f_\gamma \in L_\gamma$ $(\gamma\in\Gamma)$. In this case, if the underlying space X is a complex separable Hilbert space, every 'selection' $\{f_\gamma\}_\Gamma$ is similar to an orthonormal system [11, Theorem 3] and the constraint that Δ be finite can be dropped from the statement of Theorem 6.1.

Before giving the next theorem we prove a simple lemma which, incidentally, gives a new way of constructing ABSL's.

LEMMA 6.3. *If* H *is a (real or complex) Hilbert space and the operator* $A \in \mathcal{B}(H)$ *is injective with dense range, then* $\{L_1, L_2, L_3\}$ *where*
$L_1 = \{(x,0,0) : x \in H\}$, $L_2 = \{(x,Ax,0) : x \in H\}$ *and* $L_3 = \{(x,Ax,Ax) : x \in H\}$, *is the set of atoms of an* ABSL *on* $H \oplus H \oplus H$.

PROOF: Note that L_i $(i=1,2,3)$ is a subspace of $H \oplus H \oplus H$. Using the density of the range of A it is easily verified that

$$L_1 \vee L_2 = \{(x,y,0) : x,y \in H\},$$

$$L_2 \vee L_3 = \{(x,Ax,y) : x,y \in H\},$$

and

$$L_3 \vee L_1 = \{(x,y,y) : x,y \in H\},$$

and that $L_1 \vee L_2 \vee L_3 = H \oplus H \oplus H$.

To prove the theorem it is now enough, by Theorem 2.4, to show that

$$L_1 \cap L_2 = L_2 \cap L_3 = L_3 \cap L_1 = (0)$$

and

$$(L_1 \vee L_2) \cap (L_1 \vee L_3) = L_1, \quad (L_1 \vee L_2) \cap (L_2 \vee L_3) = L_2, \quad (L_1 \vee L_3) \cap (L_2 \vee L_3) = L_3.$$

All these proofs are easy and we do, for example, just the last one. If $(x,y,z) \in (L_1 \vee L_3) \cap (L_2 \vee L_3)$, then (x,y,z) is simultaneously of the form (p,q,q) and (r,Ar,s). So $x = p = r$, $y = q = Ar = Ax$ and $z = q = s = Ax$. Hence $(x,y,z) = (x,Ax,Ax) \in L_3$. Thus $(L_1 \vee L_3) \cap (L_2 \vee L_3) \subseteq L_3$ and reverse inclusion is obvious. ■

We are now in a position to prove the main result of this section.

THEOREM 6.4. *In every (real or complex) infinite-dimensional Hilbert space* H *there exist six subspaces* $K_i, L_i (i=1,2,3)$ *of* H *such that*

(i) $\{L_1, L_2, L_3\}$ *is the set of atoms of an* ABSL *on* H,

(ii) $0 \neq K_i \subseteq L_i$ $(i = 1,2,3)$,

(iii) $\{K_1, K_2, K_3\}$ *is not the set of atoms of an* ABSL *on* $\bigvee_1^3 K_i$.

PROOF: By writing $H = H_1 \oplus H_2$ where H_1 is infinite-dimensional and separable and by replacing $K_1, K_2, K_3, L_1, L_2, L_3 \subseteq H_1$, if necessary, by $K_1 \oplus H_2$, $K_2 \oplus (0)$, $K_3 \oplus (0)$ and $L_1 \oplus H_2$, $L_2 \oplus (0)$ and $L_3 \oplus (0)$ respectively, we may assume that H is separable. We may further assume that $H = (\ell^2)^{(6)}$, the Hilbert direct sum of six copies of ℓ^2. Let S and T be the operators on ℓ^2 defined by

$$S\,(\xi_1, \xi_2, \xi_3, \dots) = (\xi_1 - \xi_2, \frac{\xi_2 - \xi_3}{2}, \frac{\xi_3 - \xi_4}{3}, \dots)$$

and

$$T\,(\xi_1, \xi_2, \xi_3, \dots) = (\xi_1, \xi_1 - \xi_2, \frac{\xi_2 - \xi_3}{2}, \frac{\xi_3 - \xi_4}{3}, \dots).$$

Let $A \in \mathcal{B}\,(\ell^2 \oplus \ell^2)$ be defined by

$$A = \begin{bmatrix} S & 0 \\ -T & I \end{bmatrix}.$$

It is easy to see that S is injective and has dense range (because

$$S^*\,(\xi_1, \xi_2, \xi_3, \dots) = (\xi_1, -\xi_1 + \frac{1}{2}\xi_2, -\frac{1}{2}\xi_2 + \frac{1}{3}\xi_3, \dots)$$

is also injective). Hence A is injective with dense range. Define subspaces L_i $(i = 1,2,3)$ by

$$L_1 = \{(x,0,0) : x \in \ell^2 \oplus \ell^2\},$$
$$L_2 = \{(x,Ax,0) : x \in \ell^2 \oplus \ell^2\},$$
$$\text{and} \quad L_3 = \{(x,Ax,Ax) : x \in \ell^2 \oplus \ell^2\}.$$

By the preceding lemma, (i) holds. Define linear manifolds K_i $(i = 1,2,3)$ by

$$K_1 = \{(x,0,0) : x \in \ell^2 \oplus (0)\},$$
$$K_2 = \{(x,Ax,0) : x \in G(T)\},$$
$$\text{and} \quad K_3 = \{(x,Ax,Ax) : x \in G(S+T)\},$$

where for any operator B, $G(B)$ denotes the graph of B.

As $G(T)$ and $G(S+T)$ are closed, each K_i is a subspace. Clearly (ii) holds. We show that (iii) holds by showing that $(K_1 \vee K_2) \cap (K_1 \vee K_3) \neq K_1$ and appealing to Theorem 2.4. In fact we show that the vector $f = ((0,e_1), (0,0), (0,0))$ of H, where $e_1 = (1,0,0,\ldots) \in \ell^2$, belongs to $(K_1 \vee K_2) \cap (K_1 \vee K_3)$ but not to K_1.

For $n \in \mathbb{Z}^+$ define $x_n \in \ell^2$ by $x_n = \sum_{j=1}^{n} e_j$ where $\{e_j\}_1^\infty$ is the usual orthonormal basis of ℓ^2. Clearly $\lim_n Tx_n = e_1$ and $\lim_n Sx_n = 0$. Now, as $A(x_n, Tx_n) = (Sx_n, 0)$, we have

$$((-x_n,0), (0,0), (0,0)) + ((x_n,Tx_n), A(x_n,Tx_n), (0,0))$$
$$= ((0,Tx_n), (Sx_n,0), (0,0))$$
$$\to ((0,e_1), (0,0), (0,0)) = f.$$

But, for every n, the left hand side is a vector in $K_1 + K_2$ so $f \in K_1 \vee K_2$. Similarly, as $A(x_n,(S+T)x_n) = (Sx_n,Sx_n)$ we have

$$((-x_n,0),(0,0),(0,0)) + ((x_n,(S+T)x_n), A(x_n,(S+T)x_n), A(x_n,(S+T)x_n))$$
$$= ((0,(S+T)x_n), (Sx_n,Sx_n), (Sx_n,Sx_n))$$
$$\to f$$

so, as before, $f \in K_1 \vee K_3$. Hence $f \in (K_1 \vee K_2) \cap (K_1 \vee K_3)$, but clearly $f \notin K_1$. This completes the proof. ∎

We remark that, with S and T as above,

$$\mathcal{M} = \{(0), G(0), G(S), G(S+T), \ell^2 \oplus \ell^2\}$$

is a medial subspace lattice on $\ell^2 \oplus \ell^2$; that is, $M \cap N = (0)$ and $M \vee N = \ell^2 \oplus \ell^2$ for every pair M, N of distinct non-zero elements of $\mathcal{M}$ both different from $\ell^2 \oplus \ell^2$. This may provide some insight into why the above example 'works'.

7. The double commutant

Von Neumann's celebrated double commutant theorem states that a weak operator closed $*$-subalgebra of the set of operators on a complex Hilbert space, containing the identity operator, is equal to its double commutant. For example, if $\mathcal{L}$ is an ABSL with mutually orthogonal atoms, it is easy to see that Alg $\mathcal{L}$ is self-adjoint. As it is also weak operator closed, and contains the identity operator, we have $(\text{Alg } \mathcal{L})'' = \text{Alg } \mathcal{L}$.

In the past few years there has been increasing interest in non-self-adjoint algebras of operators. The ABSL's we study are defined on Banach spaces so self-adjointness (of their Alg's) is ruled out. But even if we restrict ourselves to Hilbert spaces, Alg $\mathcal{L}$ is not in general self-adjoint (it is so if and only if $M^{\perp} = M'$ for every $M \in \mathcal{L}$; equivalently, for each $M \in \mathcal{L}$ we also have $M^{\perp} \in \mathcal{L}$) . It may still happen, for an ABSL $\mathcal{L}$ with Alg $\mathcal{L}$ non-self-adjoint, that $(\text{Alg } \mathcal{L})'' = \text{Alg } \mathcal{L}$. For example the ABSL $\mathcal{L} = \{(0),L,M,X\}$ with $L + M$ a closed (rather than just dense) sum satisfies $(\text{Alg } \mathcal{L})'' = \text{Alg } \mathcal{L}$ (see [15]). On the other hand, if $L + M$ is a non-closed sum (equivalently, if the angle between L and M is zero) we have the strict inclusion $\text{Alg } \mathcal{L} \subset (\text{Alg } \mathcal{L})'' = \mathcal{B}(X)$ (see [15]).

The above remarks seem to suggest that for an ABSL $\mathcal{L}$ we have the double commutant property, $(\text{Alg } \mathcal{L})'' = \text{Alg } \mathcal{L}$, only if L and L' are at a non-zero angle for every $L \in \mathcal{L}$, $L \neq (0)$, X (the converse of this is true by Corollary 7.2 below). Perhaps surprisingly we show (Theorem 7.5) that there is an ABSL $\mathcal{L}$ on a separable Hilbert space H such that, for every $L \in \mathcal{L}$, $L \neq (0)$, H , the subspaces L and L' are at a zero angle and yet $(\text{Alg } \mathcal{L})'' = \text{Alg } \mathcal{L}$. In other words we can go to the other extreme of orthogonality of L and L' , so as far as possible from self-adjointness of Alg $\mathcal{L}$, and still retain the double commutant property. In passing we mention that the construction below gives us a new way of obtaining ABSL's.

The following is a special case of Theorem 5.4 of [15]. We shall be needing parts of the proof as well as the notation introduced therein, so we give a summary of its proof, only elaborating on the parts we shall use.

THEOREM 7.1. *Let* $\mathcal{L}$ *be an* ABSL *on a Banach space* X . *There exists a unique* ABSL $\mathcal{M}$ *on* X *such that* $(\text{Alg } \mathcal{L})'' = \text{Alg } \mathcal{M}$. *This* $\mathcal{M}$ *is a sublattice of* $\mathcal{L}$.

PROOF: As in the proof of Theorem 5.1 of [15], if $T \in (\text{Alg } \mathcal{L})'$, then the restriction of T to each atom of $\mathcal{L}$ is a multiple of the identity, equivalently, each atom of $\mathcal{L}$ is contained in an eigenspace of T . We show that in fact the eigenspaces of T are elements of $\mathcal{L}$ (and so are the closed linear spans of the atoms of $\mathcal{L}$ they contain). Let M be an eigenspace of T and let λ be the corresponding eigenvalue. Then, for every operator $A \in \text{Alg } \mathcal{L}$ and every vector $x \in M$ we have

$$TAx = ATx = \lambda Ax ,$$

so $Ax \in M$. This shows that M is invariant under every operator of Alg $\mathcal{L}$, that is $M \in \text{Lat Alg } \mathcal{L}$. But by a result of Halmos [9] , $\mathcal{L}$ is reflexive so $M \in \mathcal{L}$. (Actually the result in [9] is proved for Hilbert spaces and so is its generalization in [19], but the proof in [19] can easily be modified for Banach spaces.)

Now define the ABSL $\mathcal{M}$ as follows. For each atom $L \in \mathcal{L}$ and each operator $T \in (\text{Alg } \mathcal{L})'$ let L_T be the (unique) eigenspace of T containing L. Put

$$L_{\mathcal{M}} = \cap \{L_T : T \in (\text{Alg } \mathcal{L})'\} .$$

Clearly $L \subseteq L_{\mathcal{M}}$ and since $\mathcal{L}$ is complete, $L_{\mathcal{M}} \in \mathcal{L}$. If K is another atom of $\mathcal{L}$, either $L_T \cap K_T = (0)$ for some operator $T \in (\text{Alg } \mathcal{L})'$ in which case $L_{\mathcal{M}} \cap K_{\mathcal{M}} = (0)$, or $L_T = K_T$ for every operator $T \in (\text{Alg } \mathcal{L})'$ in which case $L_{\mathcal{M}} = K_{\mathcal{M}}$. By Example 2.7 (1)(ii) the family of subspaces $\{L_{\mathcal{M}} : L \text{ is an atom of } \mathcal{L}\}$ is the set of atoms of an ABSL, $\mathcal{M}$ say, on X. It turns out (see [15], we omit the details here) that this $\mathcal{M}$ satisfies the properties stated, moreover its uniqueness follows from the reflexivity of ABSL's. This completes the summary of the proof. ∎

COROLLARY 7.2. *Let* $\mathcal{L}$ *be an* ABSL *on a Banach space* X. *If* $L + L' = X$ *for every atom* L *of* $\mathcal{L}$ *with at most one exception (that is, there is at most one atom* M *of* $\mathcal{L}$ *with* $X = \overline{M + M'} \supset M + M'$), *then* $(\text{Alg } \mathcal{L})'' = \text{Alg } \mathcal{L}$.

PROOF: If $L + L' = X$ and L is an atom of $\mathcal{L}$, then the projection Q_L onto L along L' exists and is bounded. This projection belongs to $(\text{Alg } \mathcal{L})'$ and has L as its eigenspace corresponding to eigenvalue 1. So, if there are no exceptions, the preceding proof gives $\mathcal{M} = \mathcal{L}$.

If M is the exceptional atom, then for every other atom L of $\mathcal{L}$ we have $M \subseteq L'$. So the projection Q_L has M as part (but not all) of its eigenspace corresponding to eigenvalue 0. Since by Corollary 2.2 we have

$$\cap \{L' : L \text{ is an atom of } \mathcal{L} \text{ and } L \neq M\} = M ,$$

the result follows by the proof of Theorem 7.1. ∎

The following example shows that there can be precisely one

exceptional atom M satisfying $M + M' \neq X$. In it a gliding hump argument is employed.

EXAMPLE 7.3.

Let $\{f_n\}_1^\infty$ be a strong M-basis but not a basis of a Banach space X. In Section 5 we constructed specific such sequences $\{f_n\}_1^\infty$. We now construct an increasing sequence $\{n_k\}_1^\infty$ of positive integers as follows. Let $n_1 = 1$. Having constructed n_k we proceed by induction to construct integers m_k and n_{k+1} satisfying $n_k < m_k < n_{k+1}$ in the following manner.

Consider the Banach space $X_k = \vee\{f_n : n \geq n_k+1\}$. Clearly $\{f_n : n \geq n_k+1\}$ is a strong M-basis but not a basis of X_k. So on X_k the natural projections, each one onto the linear span of $\{f_{n_k+1}, \ldots, f_{m-1}, f_m\}$ along the closed linear span of $\{f_n : n > m\}$ for some $m \geq n_k+1$, are not uniformly bounded ([28, Vol. 1, p.25]). Thus there exists an integer $m_k \geq n_k + 1$ and such a projection P_k onto the linear span of $\{f_{n_k+1}, \ldots, f_{m_k}\}$ with $\|P_k\| \geq k + 1$. So there exist vectors

$$x_k \in \vee\{f_n : n_k + 1 \leq n \leq m_k\} \text{ and } y_k \in \vee\{f_n : n \geq m_k + 1\}$$

with $\| x_k + y_k \| < 1$ and $\| x_k \| = \|P_k(x_k + y_k)\| > k$. We may suppose that y_k belongs to the linear span of $\{f_n : n \geq m_k + 1\}$ (by perturbing y_k if necessary). Let $n_{k+1} > m_k$ be such that y_k belongs to the linear span of $\{f_n : m_k+1 \leq n \leq n_{k+1}\}$. This completes the inductive step.

Now define

$$M = \bigvee_{i=1}^{\infty} (< f_{n_i+1}, \ldots, f_{m_i} >)$$

and let $\{g_n\}_1^\infty$ be a renaming (preserving order) of the f_n's not appearing in M, that is, of

$$f_{n_1}, f_{m_1+1}, \ldots, f_{n_2}, f_{m_2+1}, \ldots, f_{n_3}, \ldots .$$

Since the subspaces $\{\langle f_n\rangle\}_1^\infty$ form the set of atoms of an ABSL (Theorem 5.1) on X, so do $\{M\} \cup \{\langle g_n\rangle : n \in \mathbb{Z}^+\}$. In this new ABSL we have $M' = \bigvee_1^\infty g_n$, moreover M and M' are at zero angle. Indeed, for every $k \in \mathbb{Z}^+$, $x_k \in M$, $y_k \in M'$, $\| x_k + y_k \| < 1$ and $\| x_k \| > k$, showing the unboundedness of the projection onto M along M'.

To summarize, we have constructed an ABSL on X with atoms M and the one-dimensional subspaces $M_n = \langle g_n\rangle$ $(n \in \mathbb{Z}^+)$, such that $M + M' \neq X$, yet for every other atom M_n we have $M_n + M_n' = X$.

For a specific instance of the above situation one may take for example $f_n = \sum_{i=1}^{n} e_i$ in ℓ^2 (see Theorems 5.7, 5.8), and put $M = \bigvee_{i=1}^{\infty} f_{2i-1}$ and $M_n = \langle f_{2n}\rangle$ $(n \in \mathbb{Z}^+)$. Here $M' = \bigvee_{i=1}^{\infty} f_{2i}$ and one can show that the angle between M and M' is zero; for example, it can easily be shown that

$$\lim_n \left\| \frac{f_{2n}}{\|f_{2n}\|} - \frac{f_{2n+1}}{\|f_{2n+1}\|} \right\|^2 = \lim_n \left(2n \left(\frac{1}{\sqrt{2n+1}} - \frac{1}{\sqrt{2n}} \right)^2 + \frac{1}{2n+1} \right) = 0 .$$

In the proof of the next theorem we shall be needing the following observation.

LEMMA 7.4. *If* I *and* J *are non-empty disjoint sets with* $I \cup J = \mathbb{Q} \cap [0,1]$, *then there exist sequences* $\{x_n\}_1^\infty$ *and* $\{y_n\}_1^\infty$ *with* $x_n \in I$, $y_n \in J$ $(n \in \mathbb{Z}^+)$ *such that* $\lim_n x_n = \lim_n y_n$.

PROOF: The hypothesis $I \cup J = \mathbb{Q} \cap [0,1]$ gives $\bar{I} \cup \bar{J} = [0,1]$. The connectedness of $[0,1]$ gives $\bar{I} \cap \bar{J} \neq \phi$. Let $x \in \bar{I} \cap \bar{J}$. Then we can find sequences $(x_n)_1^\infty$ and $(y_n)_1^\infty$ with $x_n \in I$, $y_n \in J$ $(n \in \mathbb{Z}^+)$ such that $\lim_n x_n = x = \lim_n y_n$. ■

THEOREM 7.5. *There exists an* ABSL $\mathcal{L}$ *on a separable Hilbert space* H *such that for every* $L \in \mathcal{L}$, $L \neq (0)$, H *the angle between* L *and* L' *is zero and such that* $(\mathrm{Alg}\ \mathcal{L})'' = \mathrm{Alg}\ \mathcal{L}$.

PROOF: In the following, the field of scalars can be taken to be $\mathbb{R}$ or $\mathbb{C}$. Let Q be the set of all two-element sets $\{p,q\}$ of rationals p,q with $0 \le p, q \le 1$ and $p \neq q$, that is

$$Q = \{\{p,q\} : p,q \in \mathbb{Q} \cap [0,1] \text{ and } p \neq q\} .$$

For $p \in \mathbb{Q} \cap [0,1]$ put

$$Q_p = \{\{p,q\} : q \in \mathbb{Q} \cap [0,1] \text{ and } p \neq q\} .$$

Notice that $\cup\{Q_p : p \in \mathbb{Q} \cap [0,1]\} = Q$ and that if $r \neq s$ the intersection $Q_r \cap Q_s$ consists of exactly one element $\{r,s\}$.

Our Hilbert space H will be a certain subspace of

$$\mathcal{H} = \bigoplus_{\{p,q\} \in Q} L_{\{p,q\}}$$

where $L_{\{p,q\}} = L^2[0,2\pi]$, for every $\{p,q\} \in Q$. For every $r \in \mathbb{Q} \cap [0,1]$ define L_r as the closed linear span of all those vectors of $\mathcal{H}$ with precisely one non-zero coordinate, its value being $\cos(\cdot+r)$ occuring at a coordinate $\{p,q\}$ with $r \in \{p,q\}$, that is, at a coordinate belonging to Q_r. (The functions $\cos(\cdot+r)$ could be replaced by a plethora of other families of functions. All that is required of them is that they are linearly independent and depend continuously on the parameter r.)

The Hilbert space H is now defined by $H = \vee\{L_r : r \in \mathbb{Q} \cap [0,1]\}$. A crucial observation is that H intersects each coordinate space two-dimensionally (for example, H intersects the coordinate space $L_{\{1/2,1/3\}}$ in the set of functions f of the form $f(x) = a\cos(x + 1/2) + b\cos(x + 1/3)$). To show that $\{L_r : r \in \mathbb{Q} \cap [0,1]\}$ is the set of atoms of an ABSL on H we appeal to Theorem 2.4 and show that for every pair I, J of subsets of $\mathbb{Q} \cap [0,1]$ we have

$$(\vee_I L_i) \cap (\vee_J L_j) \subseteq \vee_{I \cap J} L_k ,$$

the reverse inclusion, as usual, being obvious. We may suppose that neither I nor J is empty. Let $f \in (\vee_I L_i) \cap (\vee_J L_j)$. As $f \in \vee_I L_i$ the coordinate $f_{\{r,s\}}$ of f has one of the following forms

$$f_{\{r,s\}}(x) = \begin{cases} a \cos(x + r) + b \cos(x + s) & , \text{ if } r,s \in I, \\ a \cos(x + r) & , \text{ if } r \in I,\ s \notin I , \\ b \cos(x + s) & , \text{ if } r \notin I,\ s \in I , \\ 0 & , \text{ if } r,s \notin I , \end{cases}$$

with a and b scalars. Similar expressions hold for J , using $f \in \vee_J L_j$. As the functions $\cos(\cdot+r)$ and $\cos(\cdot+s)$ are linearly independent for $r, s \in \mathbb{Q} \cap [0,1]$, $r \neq s$, a comparison of the expressions for $f_{\{r,s\}}$, considering first I then J , gives us that in fact

$$f_{\{r,s\}}(x) = \begin{cases} a \cos(x + r) + b \cos(x + s) & , \text{ if } r,s \in I \cap J , \\ a \cos(x + r) & , \text{ if } r \in I \cap J ,\ s \notin I \cap J , \\ b \cos(x + s) & , \text{ if } r \notin I \cap J,\ s \in I \cap J , \\ 0 & , \text{ if } r,s \notin I \cap J . \end{cases}$$

(To illustrate this let us examine for example the case $r \in I \cap J$, $s \in I$, $s \notin J$. Here $f_{\{r,s\}}(x)$ equals $a \cos(x+r) + b \cos(x+s)$ and at the same time equals $c \cos(x+r)$. So $a = c$ and $b = 0$, fitting the description $a \cos(x+r)$ if $r \in I \cap J$ and $s \notin I \cap J$ given.) As $f_{\{r,s\}}$ belongs to the intersection of $\vee_{I \cap J} L_k$ with the coordinate space

$L_{\{r,s\}}$ and as f belongs to the closed linear span of its coordinates, we have $f \in \vee_{I \cap J} L_k$, as required. Hence $\{L_r : r \in \mathbb{Q} \cap [0,1]\}$ is the set of atoms of an ABSL $\mathcal{L}$ on H .

Next we show that for every proper, non-empty subset I of $\mathbb{Q} \cap [0,1]$ the subspace $M_I = \vee_I L_r$ and its Boolean complement $M'_I = \vee\{L_r : r \in (\mathbb{Q} \cap [0,1]) \setminus I\}$ are at zero angle. Indeed, for $p \in I$ and $q \notin I$ we have

$$\cos [\text{angle } (M_I, M'_I)] \geq \frac{(\cos (\cdot+p) \mid \cos (\cdot+q))}{\|\cos (\cdot+p)\| \|\cos (\cdot+q)\|}$$

$$= \frac{\int_0^{2\pi} \cos (x+p) \cos (x+q)\, dx}{(\int_0^{2\pi} \cos^2(x+p) dx)^{1/2} (\int_0^{2\pi} \cos^2(x+q) dx)^{1/2}}$$

$$= \cos (p-q) .$$

But, by Lemma 7.4, there exist sequences $\{p_n\}_1^\infty \subseteq I$ and $\{q_n\}_1^\infty \subseteq (\mathbb{Q} \cap [0,1])\setminus I$ such that $\lim_n p_n = \lim_n q_n$. Thus

$$1 \geq \cos [\text{angle } (M_I, M'_I)] \geq \lim_n \cos (p_n - q_n) = 1 ,$$

so angle$(M_I, M'_I) = 0$, as required.

Finally we show that $(\text{Alg } \mathcal{L})'' = \text{Alg } \mathcal{L}$. Let T be the (unique) linear transformation with domain the linear span of $\{L_r : r \in \mathbb{Q} \cap [0,1]\}$ satisfying $Tg = rg$ ($g \in L_r$) . We show that T is continuous on its domain (which is dense in H). For this it is enough to show, since the coordinate spaces are mutually orthogonal, that the norms of the restrictions of T to the coordinate spaces are uniformly bounded. In fact we show that on the $\{r,s\}$ coordinate space

$$\| T (a \cos (\cdot+r) + b \cos (\cdot+s)) \|^2 \leq 4 \| a \cos (\cdot+r) + b \cos (\cdot+s)\|^2 \quad (*)$$

for all (possibly complex) scalars a and b . This is equivalent to showing that

$$a^2r^2 + 2abrs\cos(r-s) + b^2s^2 \le 4\,(a^2 + 2ab\cos(r-s) + b^2)$$

for all real scalars a and b. Rearranging, we have to show that

$$(4-r^2)a^2 + 2ab\,(4-rs)\cos(r-s) + (4-s^2)b^2 \ge 0 \,.$$

This is a quadratic form in a and b with the coefficient of a^2 positive (recall that $0 \le r,s \le 1$) so it is sufficient to show that its discriminant Δ is non-positive. Here

$$\begin{aligned} \frac{1}{4}\Delta &= (4-rs)^2\cos^2(r-s) - (4-r^2)(4-s^2) \\ &= 4\,(r-s)^2 - (4-rs)^2\sin^2(r-s) \\ &\le 4\,(r-s)^2 - 3^2\sin^2(r-s)\,. \end{aligned}$$

But (with the obvious meaning when $\theta = 0$) $\sin\theta/\theta$ is decreasing on $[0,1]$, so $\sin\theta/\theta \ge \sin 1 \ge 2/3$ for $0 \le \theta \le 1$. Hence $4\theta^2 - 9\sin^2\theta \le 0$ ($\theta \in [0,1]$) and taking $\theta = |r-s|$ we get $\frac{1}{4}\Delta \le 0$, as required. (The authors thank A. Lewis for ideas leading to this simple proof of inequality (*).)

The continuity of T on its domain enables us to extend it to the whole of H without spoiling continuity. If we denote this unique extension also by T, we show that $T \in (\mathrm{Alg}\,\mathcal{L})'$. Indeed, if $r \in \mathbb{Q} \cap [0,1]$, $f \in L_r$ and $A \in \mathrm{Alg}\,\mathcal{L}$ we have $Af \in L_r$, so $TAf = rAf = ATf$. As the latter is true for every $f \in L_r$ and for every r, and as $\vee\{L_r : r \in \mathbb{Q} \cap [0,1]\} = H$, we have $AT = TA$.

We are now in a position to show that $(\mathrm{Alg}\,\mathcal{L})'' = \mathrm{Alg}\,\mathcal{L}$. By Theorem 7.1 there is a unique ABSL $\mathcal{M}$ on H with $(\mathrm{Alg}\,\mathcal{L})'' = \mathrm{Alg}\,\mathcal{M}$, so we only have to show that $\mathcal{M} = \mathcal{L}$.

The proof of Theorem 7.1 shows that every eigenspace of the specific operator T, constructed above, belongs to $\mathcal{L}$ and so is the closed linear span of the atoms it contains. It follows that, for every $r \in \mathbb{Q} \cap [0,1]$,

L_r is an eigenspace of T. Hence, since $T \in (\mathrm{Alg}\,\mathcal{L})'$, $L_r = (L_r)_{\mathcal{M}}$ for every $r \in \mathbb{Q} \cap [0,1]$ (with notation as in the proof of Theorem 7.1). It now follows that $\mathcal{M} = \mathcal{L}$ and the proof is complete. ∎

A pertinent problem is to find necessary and sufficient conditions for an ABSL $\mathcal{L}$ to satisfy $(\mathrm{Alg}\,\mathcal{L})'' = \mathrm{Alg}\,\mathcal{L}$. The above theorem and what follows shows that this is quite a difficult problem. Notice that Corollary 7.2 gives a sufficient condition which is not necessary, by Theorem 7.5. A necessary condition (Theorem 7.6(2)) is given below but this condition is not sufficient. It is however also sufficient for finite ABSL's.

THEOREM 7.6. *Let* $\mathcal{L}$ *be an* ABSL *on a Banach space* X *and let* $\mathcal{M}$ *be the unique* ABSL *on* X *with* $(\mathrm{Alg}\mathcal{L})'' = \mathrm{Alg}\mathcal{M}$.

(1) *For each finite set* $L_1, L_2, \ldots, L_n$ *of atoms of* $\mathcal{M}$, $L_1 + L_2 + \ldots + L_n$ *is a closed sum,*

(2) *If* $(\mathrm{Alg}\,\mathcal{L})'' = \mathrm{Alg}\,\mathcal{L}$, *every finite sum of atoms of* $\mathcal{L}$ *is closed.*

PROOF: The second part follows immediately from the first. For the proof of the first part we proceed by induction. For $n = 2$ let K and L be distinct atoms of $\mathcal{M}$. As $K \cap L = (0)$ the proof of Theorem 7.1 shows that there exists an operator $T \in (\mathrm{Alg}\,\mathcal{L})'$ and distinct scalars λ, μ such that $K \subseteq \ker(T - \mu I)$ and $L \subseteq \ker(T - \lambda I)$. Consider the operator $P = (T-\lambda I)/(\mu-\lambda)$ of $(\mathrm{Alg}\,\mathcal{L})'$. We have $P|_K = I$ and $P|_L = 0$, so the sum $K + L$ is closed.

For the induction step, let $K_1, K_2, \ldots, K_{n+1}$ be distinct atoms of $\mathcal{M}$. By the above, considering the pairs (K_1, K_i) $(2 \le i \le n + 1)$ there exist operators $P_i \in (\text{Alg } \mathcal{L})'$ such that

$$P_i|_{K_1} = I \quad \text{and} \quad P_i|_{K_i} = 0 \ , \quad \text{for} \quad 2 \le i \le n+1 \ .$$

Put $Q = P_2P_3 \ldots P_{n+1}$ and observe that since $(\text{Alg } \mathcal{L})'$ is abelian (for example, by Corollary 5.3 of [15] we have $(\text{Alg } \mathcal{L})' \subseteq \text{Alg } \mathcal{L}$) this product is independent of the order of the factors. Hence $Q|_{K_i} = 0$ $(2 \le i \le n+1)$ and clearly $Q|_{K_1} = I$. This shows that the sum $K_1 + \bigvee_{i=2}^{n+1} K_i$ is closed and the induction hypothesis can be used to conclude that

$K_1 + K_2 + \ldots + K_{n+1}$ is also closed. This completes the proof. ∎

Combining Corollary 7.2 and Theorem 7.6 we get the following necessary and sufficient condition for the double commutant property for ABSL's with a finite number of atoms (equivalently, with a finite number of elements).

COROLLARY 7.7. *Let* $\mathcal{L}$ *be an* ABSL *with a finite number of atoms* $L_1, L_2, \ldots, L_n$ *on a Banach space* X . *The following are equivalent.*

(1) $(\text{Alg } \mathcal{L})'' = \text{Alg } \mathcal{L}$,

(2) $L_1 + L_2 + \ldots + L_n$ *is closed* ,

(3) *for every* $L \in \mathcal{L}$, $L + L'$ *is closed.*

PROOF: That (1) implies (2) follows from Theorem 7.6. That (3) implies (1) follows from Corollary 7.2. Suppose (2) holds. Then, if $L \in \mathcal{L}$, we have

$$X \supseteq L + L' \supseteq L_1 + L_2 + \ldots + L_n = X$$

so (3) holds. ∎

The situation in the case of an infinite number of atoms is not as easy. We give an example to show that the converse of Theorem 7.6(2) is false. This example also shows that in the statement of Corollary 7.2 we cannot replace 'at most one exception' with 'at most two exceptions'.

EXAMPLE 7.8.

Let M and $<g_n>$ $(n \in \mathbb{Z}^+)$ be as in Example 7.3 and let $\mathcal{L}$ be the ABSL on $X \oplus X$ with $M \oplus (0)$, $(0) \oplus M$ and the two-dimensional subspaces $<g_n> \oplus <g_n>$ $(n \in \mathbb{Z}^+)$ as its atoms. (That these do form an ABSL follows easily from Theorem 2.4, since $\{M\} \cup \{<g_n> : n \in \mathbb{Z}^+\}$ is the set of atoms of an ABSL on X.) Clearly, every finite sum of atoms of $\mathcal{L}$ is closed. Also, in relation to Corollary 7.2, notice that we now have 'two exceptions', namely $M \oplus (0)$ (with its Boolean complement $M' \oplus X$) and $(0) \oplus M$ (with its Boolean complement $X \oplus M'$).

We show that $(\text{Alg } \mathcal{L})'' \neq \text{Alg } \mathcal{L}$. Let $T \in (\text{Alg } \mathcal{L})'$ be arbitrary. There exist scalars λ_1, λ_2 and μ_n $(n \in \mathbb{Z}^+)$ such that

$$T|_{M\oplus(0)} = \lambda_1 I, \; T|_{(0)\oplus M} = \lambda_2 I \quad \text{and} \quad T|_{<g_n> \oplus <g_n>} = \mu_n I \; (n\in\mathbb{Z}^+) .$$

We show that $\lambda_1 = \lambda_2$. Notice that for every vector $y \in M'$ there exists a vector $z \in X$ such that $T(y \oplus 0) = z \oplus 0$ and $T(0 \oplus y) = 0 \oplus z$. Since M and M' are at a zero angle there exist sequences $\{x_n\}_1^\infty \subseteq M$, $\{y_n\}_1^\infty \subseteq M'$ of unit vectors such that $\|x_n - y_n\| \to 0$. For each $n \in \mathbb{Z}^+$ let $z_n \in X$ satisfy $T(y_n \oplus 0) = z_n \oplus 0$ and $T(0 \oplus y_n) = 0 \oplus z_n$. Since $x_n \oplus 0 - y_n \oplus 0 \to 0$, applying T gives $\lambda_1 x_n \oplus 0 - z_n \oplus 0 \to 0$, so $\lambda_1 x_n - z_n \to 0$. A similar argument using $0 \oplus x_n - 0 \oplus y_n \to 0$ gives that $\lambda_2 x_n - z_n \to 0$. Hence

$$|\lambda_1 - \lambda_2| = \| (\lambda_1 x_n - z_n) - (\lambda_2 x_n - z_n) \| \longrightarrow 0$$

so $\lambda_1 = \lambda_2$.

We have therefore proved that $M \oplus (0)$ and $(0) \oplus M$ are contained in the same eigenspace of T, for every $T \in (\mathrm{Alg}\,\mathcal{L})'$. Let $\mathcal{M}$ be the unique ABSL on $X \oplus X$ satisfying $(\mathrm{Alg}\,\mathcal{L})'' = \mathrm{Alg}\mathcal{M}$ (see Theorem 7.1). Then $\mathcal{M} \neq \mathcal{L}$ because $\mathcal{M}$ has an atom containing (actually equal to) $(M \oplus (0)) \vee ((0) \oplus M) = M \oplus M$ and, of course, no atom of $\mathcal{L}$ can have this property.

References

1. N.K. Bari, *A Treatise on Trigonometric Series*, *Vol.* I, Pergamon Press, 1964.

2. G. Birkhoff, *Lattice Theory*, Amer. Math. Soc. Coll. Publ. **25**, Providence, 1948.

3. M.S. Brodskii and G.É. Kisilevskii, Quasi-direct sums of subspaces, Funct. Anal. and Applic. 1 (1967) 322-324.

4. L. Crone, D.J. Fleming and P. Jessup, Fundamental biorthogonal sequences and K-norms on ϕ, Canad. J. Math. **23** (1971) 1040-1050.

5. J.A. Erdos, Operators of finite rank in nest algebras, J. London Math. Soc. **43** (1968) 391-397.

6. J.A. Erdos and W.E. Longstaff, Commuting families of operators of rank 1, Proc. London Math. Soc. (3) **44** (1982) 161-177.

7. T. Figiel and W.B. Johnson, The approximation property does not imply the bounded approximation property, Proc. Amer. Math. Soc. **41** (1973) 197-200.

8. P.R. Halmos, Two subspaces, Trans. Amer. Math. Soc. **144** (1969) 381-389.

9. P.R. Halmos, Reflexive lattices of subspaces, J. London Math. Soc. (2) **4** (1971) 257-263.

10. K.J. Harrison, private communication.

11. K.J. Harrison and W.E. Longstaff, Automorphic images of commutative subspace lattices, Trans. Amer. Math. Soc. (1) **296** (1986) 217-228.

12. A. Hopenwasser and R. Moore, Finite rank operators in reflexive operator algebras, J. London Math. Soc. (2) **27** (1983) 331-338.

13. W.B. Johnson, On the existence of strongly series summable Markuschevich bases in Banach spaces, Trans. Amer. Math. Soc. **157** (1971) 481-486.

14. W.B. Johnson, A complementary universal Banach conjugate Banach space and its relation to the approximation problem, Israel J. Math. **13** (1972) 301-310.

15. M.S. Lambrou, Approximants, commutants and double commutants in normed algebras, J. London Math. Soc. (2) **25** (1982) 499-512.

16. M.S. Lambrou and W.E. Longstaff, Abelian algebras and reflexive lattices, Bull. London Math. Soc. **12** (1980) 165-168.

17. C. Laurie and W.E. Longstaff, A note on rank one operators in reflexive algebras, Proc. Amer. Math. Soc. (2) **89** (1983) 293-297.

18. J. Lindenstrauss and L. Tzafriri, *Classical Banach spaces I Sequence spaces*, Springer-Verlag, 1977.

19. W.E. Longstaff, Strongly reflexive lattices, J. London Math. Soc. (2) 11 (1975) 491-498.

20. W.E. Longstaff, Operators of rank one in reflexive algebras, Canad. J. Math. **28** (1976) 19-23.

21. A.S. Markus, The problem of spectral synthesis for operators with point spectrum, Math. USSR-Izv. **4** (1970) 670-696.

22. D.E. Menshov, On the partial sums of trigonometric series, M.C. **20** (62) (1947) 197-237 (in Russian).

23. A. Plans and A. Reyes, On the geometry of sequences in Banach spaces, Arch. Math. **40** (1983) 452-458.

24. H. Radjavi and P. Rosenthal, *Invariant subspaces*, Springer-Verlag, 1973.

25. A. Reyes, On the classification of sequences in Banach spaces, Arch. Math. **43** (1984) 535-541.

26. W.H. Ruckle, Representation and series summability of complete biorthogonal sequences, Pacific J. Math. **34** (1970) 511-528.

27. W.H. Ruckle, On the classification of biorthogonal sequences, Canad. J. Math. (3) **26** (1974) 721-733.

28. I. Singer, *Bases in Banach space, 2 volumes*, Springer-Verlag, 1981.

29. A. Tarski, Sur les classes d'ensembles closes par rapport à certaines opérations élémentaires, Fund. Math. **16** (1930) 181-304.

30. P. Terenzi, Block sequences of strong M-bases in Banach spaces, Collectanea Mathematice (Barcelona) **35** (1984) 93-114.

31. P. Terenzi, Representation of the space spanned by a sequence in a Banach space, Arch. Math. **43** (1984) 448-459.

S. ARGYROS and M. LAMBROU, Department of Mathematics, University of Crete, Iraklion, Crete, Greece.

W.E. LONGSTAFF, Department of Mathematics, University of Western Australia, Nedlands, Western Australia 6009, Australia.

ADDENDUM

A recent article by D.R. Larson and W.R Wogen, *Reflexivity properties of* $T \oplus 0$ (J. Funct. Anal. 92 (1990) 448 - 467) contains an example that answers several open questions. In what follows this article will be referred to as [L & W]. Of particular interest here is that it shows that there is an ABSL with one-dimensional atoms on infinite-dimensional separable Hilbert space H that fails to have the strong rank one density property. A very brief proof of this is given in [L & W]. Here we give an expanded version of that proof for the benefit of the reader. The example involves the well-known identification of $\mathcal{B}(H)$ as the dual space of the ideal $C_1(H)$ of trace-class operators via the pairing $\langle f,A\rangle = \operatorname{tr}(fA) = \operatorname{tr}(Af)$, where $f \in C_1(H)$ and $A \in \mathcal{B}(H)$. By this identification $\mathcal{B}(H)$ inherits a weak * topology.

Let $\{e_k\}_1^\infty$ be an orthonormal basis for H. Define sequences $\{f_k\}_1^\infty$, $\{g_k\}_1^\infty$ of vectors of H by $f_1 = e_1 + 4e_2$, $f_{2k-1} = -4^{k-1}e_{2k-2} + e_{2k-1} + 4^k e_{2k}$ $(k \geq 2)$, $f_{2k} = e_{2k}$ $(k \geq 1)$ and $g_{2k-1} = e_{2k-1}$, $g_{2k} = -4^k e_{2k-1} + e_{2k} + 4^k e_{2k+1}$ $(k \geq 1)$. Clearly $\{f_k\}_1^\infty$ is complete and minimal with biorthogonal sequence $\{g_k\}_1^\infty$. Put $T_k = g_k \otimes f_k$ and $Q_k = \sum_{n=1}^{k} T_n$ $(k \geq 1)$. Then $\{T_k\}_1^\infty$ and $\{Q_k\}_1^\infty$ have the same linear span, denoted by $\mathcal{S}$. The proof of Lemma 3.2 of [L & W] shows that, for every vector $x \in H$, x belongs to the closure of $\{Sx : S \in \mathcal{S}\}$. Our Theorem 5.1 now shows that $\{\langle f_k\rangle\}_1^\infty$ is the set of atoms of an ABSL $\mathcal{L}$ on H. Of course $\mathcal{S}$ is the algebra generated by the rank one operators of Alg $\mathcal{L}$. Let h be the trace class operator defined in Lemma 3.6 of [L & W]. Then $\operatorname{tr}(Q_k h) = 0$ for every $k \geq 1$, so $\operatorname{tr}(Sh) = 0$ for every $S \in \mathcal{S}$. Hence $h \in {}^{\perp}\mathcal{S}$, the pre-annihilator of $\mathcal{S}$. But $\operatorname{tr} h = 1$ so $I \notin ({}^{\perp}\mathcal{S})^{\perp}$. By a well-known result $({}^{\perp}\mathcal{S})^{\perp}$ is the weak * closure of $\mathcal{S}$. (The strong operator closure of $\mathcal{S}$ includes $({}^{\perp}\mathcal{S})^{\perp}$ and we need to show that I does not belong to the former.) If $\mathfrak{A}$ denotes the weak operator closed algebra generated by $\mathcal{S}$ and I , the relative weak operator topology and the relative weak * topology coincide on $\mathfrak{A}$ by Lemmas 3.1 and 3.5 of [L & W]. Since $\mathcal{S} \subseteq \mathfrak{A}$ and $\mathfrak{A}$ is also weak * closed, it follows that the weak operator and weak * closures of $\mathcal{S}$ coincide. Thus the identity operator does not belong to the weak operator closure of $\mathcal{S}$. The latter equals the strong operator closure by convexity. Hence the ABSL $\mathcal{L}$ does not have the strong rank one density property.

It is raised as an open question in [6] whether the converse of the result stated as Corollary 5.5 holds. The example discussed above shows that it does not. It also shows that the conclusion of Theorem 5.6 of [6] may hold even if the strong operator closed algebra generated by the given commuting family of rank one operators is not maximal abelian.

MEMOIRS of the American Mathematical Society

SUBMISSION. This journal is designed particularly for long research papers (and groups of cognate papers) in pure and applied mathematics. The papers, in general, are longer than those in the TRANSACTIONS of the American Mathematical Society, with which it shares an editorial committee. Mathematical papers intended for publication in the Memoirs should be addressed to one of the editors:

General instructions to authors for

PREPARING REPRODUCTION COPY FOR MEMOIRS

> **For more detailed instructions send for AMS booklet, "A Guide for Authors of Memoirs." Write to Editorial Offices, American Mathematical Society, P.O. Box 6248, Providence, R.I. 02940.**

MEMOIRS are printed by photo-offset from camera copy fully prepared by the author. This means that the finished book will look exactly like the copy submitted. Thus the author will want to use a good quality typewriter with a new, medium-inked black ribbon, and submit clean copy on the appropriate model paper.

Model Paper, provided at no cost by the AMS, is paper marked with blue lines that confine the copy to the appropriate size.

Special Characters may be filled in carefully freehand, using dense black ink, or **INSTANT** ("rub-on") **LETTERING** may be used. These may be available at a local art supply store.

Diagrams may be drawn in black ink either directly on the model sheet, or on a separate sheet and pasted with rubber cement into spaces left for them in the text. Ballpoint pen is not acceptable.

Page Headings (Running Heads) should be centered, in CAPITAL LETTERS (preferably), at the top of the page — just above the blue line and touching it.

LEFT-hand, EVEN-numbered pages should be headed with the AUTHOR'S NAME;

RIGHT-hand, ODD-numbered pages should be headed with the TITLE of the paper (in shortened form if necessary).

Exceptions: PAGE 1 and any other page that carries a display title require NO RUNNING HEADS.

Page Numbers should be at the top of the page, on the same line with the running heads.

LEFT-hand, EVEN numbers — flush with left margin;

RIGHT-hand, ODD numbers — flush with right margin.

Exceptions: PAGE 1 and any other page that carries a display title should have page number, centered below the text, on blue line provided.

FRONT MATTER PAGES should be numbered with Roman numerals (lower case), positioned below text in same manner as described above.

MEMOIRS FORMAT

> **It is suggested that the material be arranged in pages as indicated below. Note: Starred items (*) are requirements of publication.**

Front Matter (first pages in book, preceding main body of text).

Page i — *Title, *Author's name.

Page iii — Table of contents.

Page iv — *Abstract (at least 1 sentence and at most 300 words).

Key words and phrases, if desired. (A list which covers the content of the paper adequately enough to be useful for an information retrieval system.)

**1991 Mathematics Subject Classification.* This classification represents the primary and secondary subjects of the paper, and the scheme can be found in Annual Subject Indexes of MATHEMATICAL REVIEWS beginnning in 1990.

Page 1 — Preface, introduction, or any other matter not belonging in body of text.

Footnotes: *Received by the editor date.
Support information — grants, credits, etc.

First Page Following Introduction - Chapter Title (dropped 1 inch from top line, and centered). Beginning of Text.

Last Page (at bottom) - Author's affiliation.

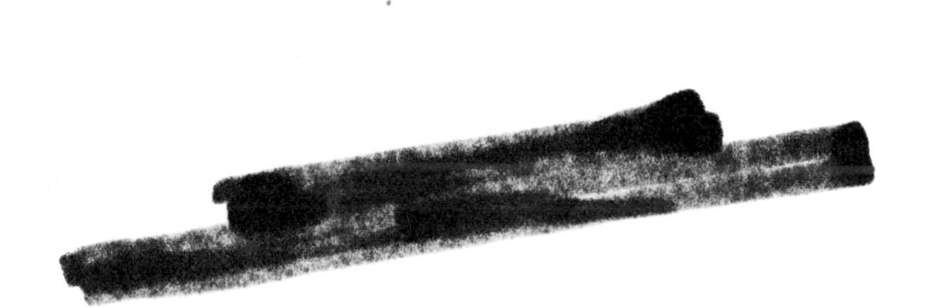